Lecture Notes in Control and Information Sciences

Volume 446

Series Editors

M. Thoma, Hannover, Germany
F. Allgöwer, Stuttgart, Germany
M. Morari, Zürich, Switzerland

Series Advisory Board

P. Fleming
University of Sheffield, UK

P. Kokotovic
University of California, Santa Barbara, CA, USA

A.B. Kurzhanski
Moscow State University, Russia

H. Kwakernaak
University of Twente, Enschede, The Netherlands

A. Rantzer
Lund Institute of Technology, Sweden

J.N. Tsitsiklis
MIT, Cambridge, MA, USA

For further volumes:
http://www.springer.com/series/642

Germán A. Ramos · Ramon Costa-Castelló,
Josep M. Olm

Digital Repetitive Control under Varying Frequency Conditions

Germán A. Ramos
Department of Electrical and Electronic
    Engineering
Universidad Nacional de Colombia
Bogotá
Colombia

Ramon Costa-Castelló
Institut d'Organització i Control de
    Sistemes Industrials (IOC)
Departament d'Enginyeria de Sistemes
    Automàtica i Informàtica Industrial
    (ESAII)
Escola Tècnica Superior d'Enginyeria
    Industrial de Barcelona (ETSEIB)
Universitat Politècnica de Catalunya (UPC)
Barcelona
Spain

Josep M. Olm
Department of Applied Mathematics IV
Universitat Politècnica de Catalunya
Castelldefels
Spain

ISSN 0170-8643 ISSN 1610-7411 (electronic)
ISBN 978-3-642-37777-8 ISBN 978-3-642-37778-5 (eBook)
DOI 10.1007/978-3-642-37778-5
Springer Heidelberg New York Dordrecht London

Library of Congress Control Number: 2013935114

© Springer-Verlag Berlin Heidelberg 2013
This work is subject to copyright. All rights are reserved by the Publisher, whether the whole or part of the material is concerned, specifically the rights of translation, reprinting, reuse of illustrations, recitation, broadcasting, reproduction on microfilms or in any other physical way, and transmission or information storage and retrieval, electronic adaptation, computer software, or by similar or dissimilar methodology now known or hereafter developed. Exempted from this legal reservation are brief excerpts in connection with reviews or scholarly analysis or material supplied specifically for the purpose of being entered and executed on a computer system, for exclusive use by the purchaser of the work. Duplication of this publication or parts thereof is permitted only under the provisions of the Copyright Law of the Publisher's location, in its current version, and permission for use must always be obtained from Springer. Permissions for use may be obtained through RightsLink at the Copyright Clearance Center. Violations are liable to prosecution under the respective Copyright Law.
The use of general descriptive names, registered names, trademarks, service marks, etc. in this publication does not imply, even in the absence of a specific statement, that such names are exempt from the relevant protective laws and regulations and therefore free for general use.
While the advice and information in this book are believed to be true and accurate at the date of publication, neither the authors nor the editors nor the publisher can accept any legal responsibility for any errors or omissions that may be made. The publisher makes no warranty, express or implied, with respect to the material contained herein.

Printed on acid-free paper

Springer is part of Springer Science+Business Media (www.springer.com)

Preface

The tracking/rejection of periodic signals constitutes a wide field of research in the control theory and applications area, and Repetitive Control has proven to be an efficient way to face this topic; however, in some applications the period of the signal to be tracked/rejected changes in time or is uncertain, which causes and important performance degradation in the standard repetitive controller. This book presents some contributions to the open topic of repetitive control working under varying frequency conditions. These contributions can be summarized as follows:

One approach that overcomes the problem of working under time varying frequency conditions is the adaptation of the controller sampling period. Nevertheless, the system framework changes from Linear Time Invariant (LTI) to Linear Time-Varying (LTV) and closed-loop stability can be compromised. This work presents two different methodologies aimed at analyzing the system stability under these conditions. The first one uses a Linear Matrix Inequality (LMI) gridding approach, which provides necessary conditions to fulfill a sufficient condition for the closed-loop Bounded Input Bounded Output (BIBO) stability of the system. The second one applies robust control techniques in order to analyze the stability and yields sufficient stability conditions. Both methodologies provide frequency variation intervals for which the system stability can be assured. Although several approaches exist for the stability analysis of general time-varying sampling period controllers, few of them allow an integrated controller design which assures closed-loop stability under such conditions. In this work two design methodologies are presented, which assure stability of the repetitive control system working under varying sampling period for a given frequency variation interval: a μ-synthesis technique and a pre-compensation strategy.

On a second branch, High Order Repetitive Control (HORC) is mainly used to improve the repetitive control performance robustness under disturbance/reference signals with varying or uncertain frequency. Unlike standard repetitive control, HORC involves a weighted sum of several signal periods. Furthermore, the use of an odd-harmonic internal model will make the system more appropriate for applications where signals have only odd-harmonic components, as in power electronics systems. Thus an odd-harmonic HORC suitable for applications involving odd-harmonic type signals with varying/uncertain frequency is presented. The open loop stability of internal models used in HORC and the one presented here is analysed. Additionally, as a consequence of this analysis, an Anti-Windup (AW) scheme for repetitive control is proposed. This AW proposal is based on the idea of having a small steady state tracking error and fast recovery once the system goes out of saturation.

The experimental validation of these proposals has been performed in two different applications: a Roto-magnet plant and an active power filter application. The Roto-magnet plant is an experimental didactic plant used as a tool for analysing and understanding the nature of the periodic disturbances, as well as to study the different control techniques used to tackle this problem. This plant has been adopted as experimental test bench for rotational machines. On the other hand, shunt active power filters have been widely used as a mean to overcome the power quality problems caused by nonlinear and reactive loads. These power electronics devices are designed with the goal of obtaining a power factor close to 1 and achieving current harmonics and reactive power compensation. The implementation on this plant shows the experimental behaviors of the here proposed repetitive controllers under constant and varying network frequency. Moreover, this performance is presented and analyzed in terms of the Total Harmonic Distortion (THD), Power Factor (PF) and $cos\phi$.

This monograph is the result of more than four years of research at the Institute of Industrial and Control Engineering of the Universitat Politècnica de Catalunya. We would like to thank the financial support received from the Spanish Ministry of Science through the projects DPI2007-62582 and DPI2010-15110.

Contents

1 **Introduction** 1
   1.1   Motivation and Problem Statement 1
   1.2   Contribution 2
   1.3   Outline 3
   References 3

2 **Repetitive Control** 5
   2.1   Basics 5
       2.1.1   The Internal Model Principle 5
       2.1.2   The Repetitive Controller 7
   2.2   Performance under Varying Frequency 10
   References 12

Part I: Varying Sampling Approach

---

3 **Stability Analysis Methods** 15
   3.1   Introduction 15
       3.1.1   State of the Art 15
       3.1.2   Contribution 16
       3.1.3   Outline 17
   3.2   Repetitive Control under Varying Frequency Conditions 17
   3.3   LMI Gridding Approach 19
   3.4   Robust Analysis 21
   3.5   Conclusions 24
   References 24

4 **Design Methods** 27
   4.1   Introduction 27
       4.1.1   State of the Art 27
       4.1.2   Contribution 28

		4.1.3	Outline	28
	4.2	Robust Stability Design		28
	4.3	Plant Pre-compensation		31
	4.4	Conclusions		33
	References			34

Part II: HORC Approach

5 Odd-Harmonic High Order Repetitive Control 37
  5.1 Introduction 37
    5.1.1 State of the Art 37
    5.1.2 Contribution 39
    5.1.3 Outline 40
  5.2 Internal Model Poles Analysis 40
  5.3 Odd-Harmonic HORC 43
    5.3.1 Odd-Harmonic Repetitive Control 43
    5.3.2 Odd-Harmonic HORC Internal Model 44
    5.3.3 Selection of the Gain k_r 47
    5.3.4 Performance under Varying Frequency Conditions 48
    5.3.5 Second-Order Odd-Harmonic Internal Model 49
  5.4 Anti-windup Synthesis for Repetitive Control 52
    5.4.1 Introduction 52
    5.4.2 The General MRAW Scheme 54
    5.4.3 MRAW Proposal: The Deadbeat Anti-windup
           Controller for RC 56
    5.4.4 MRAW Proposal: Design and Stability 58
    5.4.5 Stability 58
    5.4.6 Optimal Design 61
  5.5 Conclusions 62
  References 63

Part III: Experimental Validation

6 Roto-Magnet 67
  6.1 Plant Description 68
  6.2 Standard Repetitive Controller 69
  6.3 The Varying Sampling Period Strategy 73
    6.3.1 Implementation Issues 73
    6.3.2 LMI Gridding Approach 75
    6.3.3 Robust Control Theory Approach 76
    6.3.4 Experimental Results 78
  6.4 Robust Design 82
    6.4.1 Experimental Results 85

	6.5	Adaptive Pre-compensation	86
		6.5.1 Controller Design	86
		6.5.2 Experimental Results	88
	6.6	Anti-windup Optimal Design for HORC	89
		6.6.1 Experimental Setup	90
		6.6.2 Experimental Results	91
	6.7	Conclusions	99
	References		99
7	**Shunt Active Power Filter**		**101**
	7.1	Plant Description	102
		7.1.1 Control Objectives	103
		7.1.2 Controller Structure	104
	7.2	Odd Harmonic Repetitive Controller	106
		7.2.1 Performance at Nominal Frequency	107
		7.2.2 Performance under Network Frequency Variations	113
	7.3	Varying Sampling Results	114
		7.3.1 Implementation Issues	114
		7.3.2 Robust Control Theory Approach	115
		7.3.3 Experimental Results	116
	7.4	Adaptive Pre-compensation	119
		7.4.1 Controller Design	119
		7.4.2 Controller Calculation	120
		7.4.3 Experimental Results	123
	7.5	Robust Design	126
		7.5.1 Experimental Results	126
	7.6	HORC	129
		7.6.1 Experimental Setup	129
		7.6.2 Experimental Results	130
	7.7	Conclusions	134
	References		136
8	**Conclusions**		**139**
	8.1	Conclusions	139

Appendices

A	**Implementation of the Stabilizing Filter**	**145**
	References	146
B	**Calculation of the Sampling Period Variation Interval**	**147**
	B.1 First-Order Plants	147
	B.2 Higher Order Plants	148

	B.2.1 Numerical Calculation 148
	B.2.2 Log Norm Bound 148
	B.2.3 Schur Decomposition-Derived Bound 149

References 150

C Optimal LQ Design in LMI Form **151**

References 153

D List of Symbols **155**

Index **159**

List of Figures

2.1 Basic structure of continuous-time T_p-periodic signal generator 6
2.2 Generic repetitive control internal model scheme, where $W(z)$ is a delay function, $H(z)$ a null-phase low-pass filter, and $\sigma \in \{-1, 1\}$ 7
2.3 Discrete-time block-diagram of the proposed repetitive transfer function 8
2.4 Open-loop IM gain diagram 11
2.5 Closed-loop transfer function phase diagram 11

3.1 Accommodation of the sampling period T_s to possible variations of T_p 17
3.2 Graphical scheme of the LMI gridding-based stability analysis in $\mathcal{T} = [T_0, T_f]$ from the solution of (3.6) 20
3.3 Graphical scheme of the LMI gridding-based stability analysis in $\mathcal{T} = [T_0, T_f]$ from the solution of (3.7) 20
3.4 The resulting feedback system 22

4.1 Adaptation of the repetitive control loop to the H_∞ formulation 30
4.2 Equivalent scheme 30
4.3 Discrete-time block-diagram of the closed-loop system with the proposed repetitive controller structure 32
4.4 Detail of the compensator-plant system 32

5.1 Block-diagram of the repetitive controller plug-in approach 38
5.2 Magnitude response of $S_{Mod}(z)$: comparison of [11], [1], [19] and [15] for $G_x(z) = 1/(G_o(z))^{-1}$, $H(z) = 1$ and $M = 3$ 39
5.3 Nyquist plot of $-W(z)$ with $N = 400$ for traditional RC ($M = 1$) and HORC ($M = 3$) tuned according to [11], [1], [19] and [15] 41
5.4 Nyquist plot of $-W(z)H(z)$ for HORC ($M = 3$) tuned according to [19] and using $H(z) = 0.05z^{-1} + 0.9 + 0.05z$ 42

XII   List of Figures

5.5   Nyquist plot of $-W(z)H(z)$ for HORC ($M = 3$) tuned according to [15] and using $H(z) = 0.05z^{-1} + 0.9 + 0.05z$ 42
5.6   Sensitivity functions of generic (Steinbuch (2002)) and odd-harmonic high-order repetitive controllers with $M = 3$ and $H(z) = 1$ 46
5.7 $S_{Mod}(z)$ magnitude response for several values of k_r. Top: $M = 2$, bottom: $M = 3$. 48
5.8   Closed-loop poles modulus vs k_r for odd-harmonic RC ($M = 1$), second order ($M = 2$) and third order ($M = 3$) HORC, using $N = 100$ 49
5.9   Odd-harmonic RC and odd-harmonic HORC internal models gain diagram 50
5.10 Values of k_r vs. M for which the system is stable. $H(z)$ is a null-phase filter with $\|H(z)\|_\infty < 1$, $|H(e^{j\bar{\omega}})|_{\omega=0} = 1$ and $G_p(z)$ is a minimum-phase plant. 52
5.11 The MRAW scheme in RC 54
5.12 The invariant part of the MRAW scheme 56
5.13 AW compensator comparison: Ideal response for the system without saturation, IMC for $K = 0$, DB for a deadbeat behaviour and LQR for an intermediate behaviour 57
5.14 The connection between the AW compensator $C_{aw}(z)$ and the saturation block 59

6.1   Mechanical load: fixed and moving permanent magnets sketch (ω and Γ_p stand for the angular speed and the disturbance torque, respectively) 68
6.2   Picture of the main part of the Roto-magnet plant: DC motor, optical encoder, magnetic system (load), and supporting structure 68
6.3   Closed-loop time response of the system without the electromagnets 69
6.4   Closed-loop time response and harmonic content without repetitive controller and with the electromagnets 70
6.5   Frequency response of the open-loop function $G_l(z)$ 70
6.6   Frequency response of the sensitivity function $S(z)$ 71
6.7   Steady-state response and output signal harmonic content of the RC system 71
6.8   Step response of the designed RC system 72
6.9   First harmonic gain factor evolution at harmonic frequencies 74
6.10 Maximum eigenvalue of $L_{T_k}(P_N)$ with $\alpha = 4645$ and $\bar{T} = 0.005\text{s}$... 76
6.11 Maximum eigenvalue of $L_{T_k}(P_G)$ with 40 points for the design grid and 55121 points to check stability 77

6.12 Closed-loop system behavior using a repetitive controller
and with sampling period T_s: (a) Fixed at the nominal value
($\bar{T} = 1$ ms); (b) Obtained from a second order frequency observer
for T_p; (c) Obtained from an exact estimation of T_p 79
6.13 Closed-loop system behavior using a repetitive controller
and with sampling period T_s: (a) Fixed at the nominal value
($\bar{T} = 5$ ms); (b) Obtained from a second order frequency observer
for T_p; (c) Obtained from an exact estimation of T_p 80
6.14 Sampling period T_s corresponding to an exact (blue) and a
second order (green) observer for T_p. Design for $\bar{T} = 1$ms. 81
6.15 Sampling period T_s corresponding to an exact (blue) and a
second order (green) observer for T_p. Design for $\bar{T} = 5$ms. 81
6.16 Adaptation of the $G_o(z)$ control loop to the H_∞ formulation 82
6.17 The resulting feedback system with the uncertainty 83
6.18 Frequency response comparison between $\tilde{G}_x(z)$ and the original
controller $G_x(z) = k_r G_o^{-1}(z)$, with $k_r = 0.21$ 83
6.19 Open-loop magnitude response of function $G_l(z)$. Standard
design (SD) and robust design (RD) comparison. 84
6.20 Sensitivity function magnitude response. Standard design (SD)
and robust design (RD) comparison. 84
6.21 Steady-state time-response and harmonic content using the RC
robust design for nominal speed $\omega = 4$ rev/s 85
6.22 Closed-loop system behaviour using the RC robust design with
varying sampling rate T_s obtained from an exact estimation of T_p ... 86
6.23 Closed-loop system behaviour using a repetitive controller
with adaptive pre-compensation with varying sampling rate T_s
obtained from an exact estimation of T_p 89
6.24 Magnitude response of open-loop function $G_l(z)$. RC design and
HORC design comparison. 90
6.25 Sensitivity function magnitude response. RC design and HORC
design comparison. 91
6.26 Steady-state time-response and harmonic content of the RC for a
0.5% speed deviation 92
6.27 Steady-state time-response and harmonic content of the HORC
for a 0.5% speed deviation 93
6.28 Steady-state time-response of the system without AW
compensator 94
6.29 Detailed control action of the system without AW compensator 94
6.30 Steady-state time-response of the system with IMC AW
compensator 95
6.31 Steady-state time-response of the system with deadbeat optimal
AW compensator 95
6.32 Detailed error $e_k = r_k - y_k$ and control signal of the system with
deadbeat optimal AW compensator 96

6.33 Steady-state time-response of the system without AW compensator 96
6.34 Detailed control action of the system without AW compensator 97
6.35 Steady-state time-response of the system with IMC AW compensator 97
6.36 Steady-state time-response of the system with DB optimal AW compensator 98
6.37 Steady-state error comparison of the three AW strategies. e_1 stands for no AW compensation, e_2 for IMC AW strategy and e_3 for the optimal AW proposal. 98

7.1 Single-phase shunt active filter connected to the network-load system 102
7.2 Global architecture of the control system 104
7.3 Current control block diagram 104
7.4 Energy shaping controller block diagram 106
7.5 Magnitude response of the open-loop function $G_l(z)$ for the active filter RC design 108
7.6 Magnitude response of the sensitivity function for the active filter RC design 109
7.7 Nonlinear load connected to the AC source working at 50 Hz. Top: v_n, v_1, i_l and i_n. Bottom: PF, $\cos\varphi$ and THD for i_n. 109
7.8 Nonlinear load and the active filter connected to source (50 Hz). Top: v_n, i_n, i_l and v_1; Bottom: PF, $\cos\phi$ and THD for i_n. 110
7.9 An off-on transition of the nonlinear load with the active filter connected to source (50 Hz). v_n, i_n, v_1 and v_2. 110
7.10 Linear load and the active filter connected to source (50 Hz). Top: v_n, i_n, i_l and v_1; Bottom: PF, $\cos\phi$ and THD for i_n. 111
7.11 Nonlinear load and the active filter connected to source (49.5 Hz). Top: v_n, i_n, v_1, and v_2; Bottom: PF, $\cos\phi$ and THD for i_n. 111
7.12 THD degradation due to the frequency variation having a nonlinear load 112
7.13 $cos\phi$ and PF degradation due to the frequency variation having a nonlinear load 112
7.14 PF variation as a function of the network frequency having a pure capacitive load 113
7.15 Network frequency computation 115
7.16 Nonlinear load with active filter connected to the ac source working at 48 Hz with adaptive scheme. Top: v_n, v_1, i_l and i_n. Bottom: PF, $\cos\varphi$ and THD for i_n. 116
7.17 Nonlinear load with active filter connected to the ac source working at 52 Hz with adaptive scheme. Top: v_n, v_1, i_l and i_n. Bottom: PF, $\cos\varphi$ and THD for i_n. 117

7.18 Nonlinear load with active filter connected to the ac source with adaptive scheme. v_n and i_n when the network frequency changes from 53 Hz to 48 Hz in 49 cycles. 117
7.19 Linear load and the active filter connected to source (51 Hz). Top: v_n, i_n, i_l and v_1; Bottom: PF, $\cos\phi$ and THD for i_n. 118
7.20 Linear load and the active filter connected to source (49 Hz). Top: v_n, i_n, i_l and v_1; Bottom: PF, $\cos\phi$ and THD for i_n. 118
7.21 Discrete-time block-diagram of the basic varying sampling repetitive control structure 119
7.22 Discrete-time block-diagram of the closed-loop system with the adaptation-compensation controller structure 119
7.23 Pre-filtering scheme 121
7.24 Internal stability evaluation. Eigenvalues evolution of $L_{T_k}(P_G)$ using 40 points in the design grid and 50000 points to check stability. 122
7.25 Nonlinear load with active filter connected to the ac source working at 48 Hz with pre-compensation scheme. Top: v_n, v_1, i_l and i_n. Bottom: PF, $\cos\varphi$ and THD for i_n. 124
7.26 Nonlinear load with active filter connected to the ac source working at 53 Hz with pre-compensation scheme. Top: v_n, v_1, i_l and i_n. Bottom: PF, $\cos\varphi$ and THD for i_n. 124
7.27 Nonlinear load with active filter with pre-compensation scheme when the network frequency changes from 48Hz to 53Hz: v_n, i_n ... 125
7.28 Frequency response comparison between $\tilde{G}_x(z)$ and of the original controller $G_x(z) = k_r G_o^{-1}(z)$, with $k_r = 0.197$ 125
7.29 Open-loop magnitude response of function $G_l(z)$. Standard design (SD) and robust design (RD) comparison. 127
7.30 Sensitivity function magnitude response. Standard design (SD) and robust design (RD) comparison. 127
7.31 Nonlinear load with active filter connected to the ac source working at 49 Hz using the robust design. Top: v_n, v_1, i_l and i_n. Bottom: PF, $\cos\varphi$ and THD for i_n. 128
7.32 Nonlinear load with active filter connected to the ac source working at 51 Hz using the robust design. Top: v_n, v_1, i_l and i_n. Bottom: PF, $\cos\varphi$ and THD for i_n. 128
7.33 Nonlinear load with active filter connected to the ac source using the robust design. v_n and i_n with a step change in the network frequency from 50 Hz to 51 Hz. 129
7.34 Open-loop magnitude response of function $G_l(z)$. Odd-harmonic RC and HORC design comparison. 130
7.35 Sensitivity function magnitude response. Odd-harmonic RC and HORC design comparison. 131
7.36 Nonlinear load and the active filter connected to source (50 Hz) using the odd-harmonic HORC. (top) v_n, i_n, i_l and v_1 vs time; (bottom) PF, $\cos\phi$ and THD for i_n. 131

7.37 Nonlinear load and the active filter connected to source (49.5 Hz) using the odd-harmonic HORC. (top) v_n, i_n, i_l and v_1 vs time; (bottom) PF, $\cos\phi$ and THD for i_n. 132

7.38 Nonlinear load and the active filter connected to source (50.5 Hz) using the odd-harmonic HORC. (top) v_n, i_n, i_l and v_1 vs time; (bottom) PF, $\cos\phi$ and THD for i_n. 132

7.39 Nonlinear load and the active filter connected to source using the odd-harmonic HORC. Frequency transition from 49.5 Hz to 50.5 Hz. v_n, i_n, and v_1 vs time. 133

7.40 Linear load and the active filter connected to source using the odd-harmonic HORC (49.5 Hz). Top: v_n, i_n, i_l and v_1; Bottom: PF, $\cos\phi$ and THD for i_n. 133

7.41 Linear load and the active filter connected to source using the odd-harmonic HORC (50.5 Hz). Top : v_n, i_n, i_l and v_1; Bottom: PF, $\cos\phi$ and THD for i_n. 134

A.1 Implementation of $G_x(z)$ and the internal model. $W(z)$ is a delay function, $H(z)$ a null-phase low pass filter, and $\sigma \in \{-1, 1\}$. 145

List of Tables

2.1　Some internal models used in repetitive control ....　7

5.1　Obtained weights using the proposals [11], [1], [19] and [15] $M = 3$ ....　40

6.1　Normed error for the three proposed designs with time varying sample time: Standard, Pre-compensation and Robust ....　88

7.1　Shunt active filter: stability intervals in frequency units (Hz) ....　115

1
Introduction

Summary. Repetitive Control has proved to be an excellence technique to be applied in systems subject to periodic references/disturbances. Unfortunately, it suffers a dramatic performance decay when the signal frequency is uncertain or time varying. This chapter briefly introduces the solution proposed in this work.

1.1 Motivation and Problem Statement

Repetitive control [6, 11, 17, 19] is an IMP-based control technique [7] that yields perfect asymptotic tracking and rejection of periodic signals. Essentially, this is achieved by including a generator of the reference/disturbance signal in the control loop. Periodic signals arise in many real-world applications and repetitive control has been successfully used in different control areas, such as CD and disk arm actuators [20], robotics [16], electro-hydraulics [13], tubular heat exchangers [2], electronic rectifiers [23], pulse-width modulated inverters [21, 22] and shunt active power filters [8] among others.

It is usual to design repetitive controllers assuming a constant period T_p for the signals to be tracked/rejected. Then, a constant sampling period T_s is selected and, eventually, the value of the ratio $N = T_p/T_s$ is structurally embedded in the control algorithm. However, it is well known that even slight changes in the frequency of the tracked/rejected signals result in a dramatic decay of the controller performance [14]. A large number of practical applications can be affected by this phenomenon including disturbance period changes caused by speed variations in rotational mechanical systems [4] (CD-roms, printers, peristaltic pumps, machining tools, etc.), and frequency variations undergone by power devices (shunt active filters, inverters, rectifiers) connected to the electric distribution network [5].

Several approaches have been introduced to overcome this problem, which can be grouped in three sets:

- A first set of proposals dealing with this problem maintain the sampling period T_s constant and adapt the value of N according to the time variation of T_p [12, 18]. This renders a variable structure IM with an integer approximation N of the ratio T_p/T_s which, under the variations of T_p, is generally a non integer number.

- The following two sets of approaches work with a fixed value of N, its main advantage being quality preservation in signal reconstruction. Thus, the second set of proposals also maintain the initially selected sampling period, T_s, and robustness against frequency variations is achieved by means of large memory elements [14, 15] or introducing a fictitious sampler operating at a variable sampling rate and later using a fixed frequency IM [1, 3]. The former idea works well for small frequency variations at the cost of increasing the computational burden, while the latter constitutes a very simple method with less computational load but presents some performance degradation in the high frequency range due to the use of interpolation.
- The third set of schemes proposes an adaptation of the controller sampling rate according to the reference/disturbance period [9, 10, 18]. This allows the preservation of the steady-state performance while maintaining a low computational cost but, on the other hand, the original LTI system becomes LTV. This structural change requires a new stability study, but no formal proofs regarding this issue are reported in the quoted references.

1.2 Contribution

This work presents some contributions to the open topic of RC working under varying frequency conditions. These contributions are developed in the SISO systems framework and can be summarized as follows:

- **Stability analysis methods for repetitive control under varying sampling time**. As mentioned before, an efficient way to deal with frequency variations is to adapt the controller sampling time. However, a stability analysis should be carried out in order to establish the formal validity of this technique. Two different methodologies are presented: LMI gridding and robust stability analysis. The first technique yields necessary conditions for a sufficient stability condition and the second one renders sufficient stability conditions. In both approaches, the stability conditions are given for a sampling time variation interval, using the LMI gridding method the stability is rigorously guaranteed only in open neighborhoods of selected grid points within the interval, while the robust control method can assure stability for the entire variation interval.
- **Design methods for repetitive controllers dealing with varying frequency conditions**. In the frame of a variable sampling time strategy the design problem can be stated as finding a repetitive controller that assures the system stability and performance for a given frequency variation interval. Following this idea, two designs are proposed: a plant pre-compensation, where an additional compensator is added to the standard scheme to provide compensation for the sampling time variation, and a robust μ-synthesis design, where the part of the system affected by the sampling period variation is treated as a bounded structured uncertainty, thus obtaining a controller that stabilizes the system and preserves performance for the required frequency interval.

- **Robust performance of repetitive control using high order internal models**. High order repetitive controllers are robust against frequency variation . However, they require a higher computational load compared with the standard controller. Additionally, in some applications like active power filters one needs to reject disturbance signals with only odd harmonic components. In this sense, an odd-harmonic HORC is proposed, which is able to reject odd-harmonic components and can be implemented with a similar computational burden to that of the standard repetitive controller. Finally, due to the characteristics of the obtained internal models an AW compensation scheme is added. The design involves an optimal LQ approach to obtain the AW filter.

1.3 Outline

The book starts with two introductory chapters: Chapter 1 presents and discusses the state of the art, motivation and contributions of this work and Chapter 2 summarizes the RC concepts, basics and drawbacks. Following these chapters the book is divided in three parts.

The first part deals with the varying sampling period approach, the second part proposes a robust performance strategy and the final part presents the experimental validation of the previously described proposals. The first part is, in turn, divided in two chapters. Chapter 3 presents two stability analysis for systems that work with varying sampling period: an LMI gridding approach and a robust control theory-based analysis. In Chapter 4 two control design methods are described: a pre-compensation scheme and a robust control theory-based design. The second part is in Chapter 5, which introduces the odd-harmonic HORC and introduces an AW scheme for this type of controller. The third part includes the experimental validation through two different plants: Chapter 6 presents the experimental results of the Roto-magnet didactic plant and Chapter 7 describes an active power filter application. Finally, conclusions are outlined in Chapter 8.

References

1. Álvarez, J., Yebra, L., Berenguel, M.: Adaptive repetitive control for resonance cancellation of a distributed solar collector field. International Journal of Adaptive Control Signal Processing 23(4), 331–352 (2007)
2. Álvarez, J., Yebra, L., Berenguel, M.: Repetitive control of tubular heat exchangers. Journal of Process Control 17(9), 689–701 (2007)
3. Cao, Z., Ledwich, G.F.: Adaptive repetitive control to track variable periodic signals with fixed sampling rate. IEEE/ASME Transactions on Mechatronics 7(3), 374–384 (2002)
4. Chew, K., Tomizuka, M.: Digital control of repetitive errors in disk drive systems. IEEE Control Systems Magazine 10(1), 16–19 (1990)
5. Costa-Castelló, R., Malo, S., Griñó, R.: High performance repetitive control of an active filter under varying network frequency. In: Proceedings of the 17th IFAC World Congress, pp. 3344–3349 (2008)

6. Costa-Castelló, R., Nebot, J., Griñó, R.: Demonstration of the internal model principle by digital repetitive control of an educational laboratory plant. IEEE Transactions on Education 48(1), 73–80 (2005)
7. Francis, B., Wonham, W.: Internal model principle in control theory. Automatica 12, 457–465 (1976)
8. Griñó, R., Cardoner, R., Costa-Castelló, R., Fossas, E.: Digital repetitive control of a three-phase four-wire shunt active filter. IEEE Transactions on Industrial Electronics 54(3), 1495–1503 (2007)
9. Hanson, R.D., Tsao, T.-C.: Periodic sampling interval repetitive control and its application to variable spindle speed noncircular turning process. Journal of Dynamic Systems, Measurement, and Control 122(3), 560–566 (2000)
10. Hillerström, G.: On Repetitive Control. PhD thesis, Lulea University of Technology (November 1994)
11. Hillerström, G., Walgama, K.: Repetitive control theory and applications - A survey. In: Proceedings of the 13th IFAC World Congress, vol. D, pp. 1–6 (1996)
12. Hu, J.: Variable structure digital repetitive controller. In: Proceedings of the American Control Conference, vol. 2, pp. 2686–2690 (1992)
13. Kim, D.H., Tsao, T.: Robust performance control of electrohydraulic actuators for electronic cam motion generation. IEEE Transactions on Control Systems Technology 8(2), 220–227 (2000)
14. Steinbuch, M.: Repetitive control for systems with uncertain period-time. Automatica 38(12), 2103–2109 (2002)
15. Steinbuch, M., Weiland, S., Singh, T.: Design of noise and period-time robust high order repetitive control, with application to optical storage. Automatica 43, 2086–2095 (2007)
16. Tayebi, A., Abdul, S., Zaremba, M., Ye, Y.: Robust iterative learning control design: Application to a robot manipulator. IEEE/ASME Transactions on Mechatronics 13(5), 608–613 (2008)
17. Tomizuka, M.: Dealing with periodic disturbances in controls of mechanical systems. Annual Reviews in Control 32(2), 193–199 (2008)
18. Tsao, T.-C., Qian, Y.-X., Nemani, M.: Repetitive control for asymptotic tracking of periodic signals with an unknown period. Journal of Dynamic Systems, Measurement, and Control 122(2), 364–369 (2000)
19. Wang, Y., Gao, F., Doyle III, F.J.D.: Survey on iterative learning control, repetitive control, and run-to-run control. Journal of Process Control 19(10), 1589–1600 (2009)
20. Wu, S.-C., Tomizuka, M.: An iterative learning control design for self-servowriting in hard disk drives. Mechatronics 20(1), 53–58 (2010)
21. Zhang, B., Zhou, K., Wang, Y., Wang, D.: Performance improvement of repetitive controlled PWM inverters: A phase-lead compensation solution. International Journal of Circuit Theory and Applications 38, 453–469 (2010)
22. Zhou, K., Wang, D.: Digital repetitive learning controller for three-phase CVCF PWM inverter. IEEE Transactions on Industrial Electronics 48(4), 820–830 (2001)
23. Zhou, K., Wang, D., Xu, G.: Repetitive controlled three-phase reversible PWM rectifier. In: Proceedings of the American Control Conference, pp. 125–129 (2000)

2
Repetitive Control

Summary. This Chapter describes the concepts and basics of RC. The standard IM and the plug-in scheme structure are used to introduce the design, as well as stability and robustness approaches that are traditionally employed in this technique. The performance of the RC interface of frequency variations or uncertainty is analyzed using the magnitude response of the IM and the closed loop phase behavior of the system. These response characteristics evidence the dramatic loss of performance that occurs when the period of the reference/disturbance signal is time varying or uncertain. Section 2.1 introduces the IM, the controller structure, stability conditions and design criteria in RC, while in Section 2.2 the performance degradation under varying frequency conditions is analysed.

2.1 Basics

2.1.1 The Internal Model Principle

As an IMP-based [6] strategy, RC uses an IM which corresponds with the model of a periodic signal. In order to derive this model, recall that the trigonometric Fourier series expansion of a T_p-periodic signal $r(t)$ reads as:

$$r(t) = a_0 + \sum_{k=1}^{\infty} a_k \cos \frac{2k\pi}{T_p} t + b_k \sin \frac{2k\pi}{T_p} t. \qquad (2.1)$$

By the IMP, the inclusion of the generator of (2.1) in the control loop results in the tracking/rejection of any T_p-periodic reference/disturbance signal. Hence, following [16], the transfer function of a periodic signal generator may be written as

$$\hat{G}_r(s) = \frac{1}{s} \prod_{k=1}^{\infty} \frac{\left(\frac{2k\pi}{T_p}\right)^2}{s^2 + \left(\frac{2k\pi}{T_p}\right)^2} = \frac{T_p e^{-\frac{T_p s}{2}}}{1 - e^{-T_p s}}. \qquad (2.2)$$

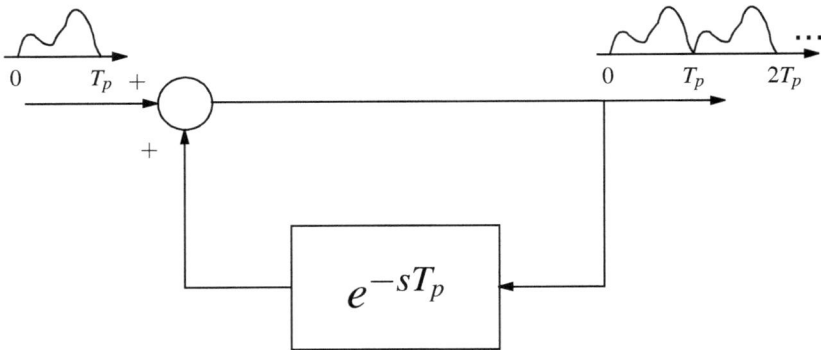

Fig. 2.1 Basic structure of continuous-time T_p-periodic signal generator

However, $T_p e^{-T_p s/2}$ being a delay term with a gain T_p, it is sufficient to include the following continuous time IM:

$$\bar{G}_r(s) = \frac{1}{1 - e^{-T_p s}}$$

inside the control loop, which can be implemented as $e^{-T_p s}$ with a positive feedback, as depicted in Figure 2.1. Notice that the IM(2.2) has poles at $s = \pm jk/T_p$, $k \in \mathbb{N}$. Therefore, from a frequency point of view, $\bar{G}_r(s)$ exhibits infinite gain at frequencies k/T_p, $\forall k \in \mathbb{N}$. This assures zero tracking error at these frequencies in closed loop if the closed-loop system is stable.

It is also worth mentioning that some studies relate RC with learning control techniques (see, for example, [16]). This is due to the fact that the basic repetitive structure learns a signal of length T_p and repeats it as a periodic signal of period T_p if the input to the system is set to zero (see Figure 2.1).

The implementation of a time delay in continuous-time is a complicated point [11]. Fortunately, in discrete time it is an easier task: if the reference/disturbance signal period T_p is a multiple of the sampling period T_s, the digital implementation is reduced to a circular queue. Therefore, the discrete IM that should be included in the loop is:

$$G_r(z) = \frac{z^{-N}}{1 - z^{-N}} = \frac{1}{z^N - 1}, \quad (2.3)$$

where $N = T_p/T_s \in \mathbb{N}$.

In addition to the constraint that represents the demand of a constant ratio between T_p and T_s, it is important to point out that T_s should be selected taking into account that discrete-time implementations can only deal with those harmonics which are below the Nyquist frequency $\omega_s/2 = \pi/T_s$. A magnitude response of the IM (2.3) designed for a 50 Hz periodic signal is depicted in Figure 2.4. It can be noticed that the response presents high gain at the fundamental and harmonic frequencies.

Several types of internal models are used depending on the specific periodic signal to deal with [3, 4, 7, 10].

2.1 Basics   7

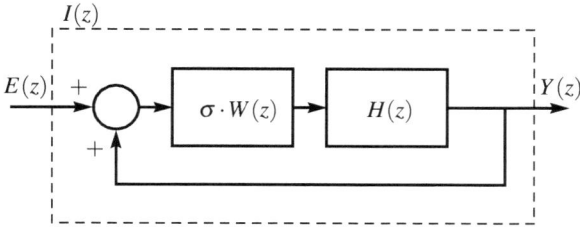

Fig. 2.2 Generic repetitive control internal model scheme, where $W(z)$ is a delay function, $H(z)$ a null-phase low-pass filter, and $\sigma \in \{-1, 1\}$

In this work the following generic IM will be used:

$$I(z) = \frac{\sigma W(z) H(z)}{1 - \sigma W(z) H(z)}, \qquad (2.4)$$

where $W(z)$ is a time delay function, $H(z)$ is a low-pass filter for robustness improvement at high frequencies [2] and σ determines the feedback sign i.e. $\sigma = 1$ and $\sigma = -1$ for positive and negative feedback respectively. Figure 2.2 shows the feedback scheme of this IM. From the generic model (2.4) the so-called standard, odd-harmonic, high order and odd-harmonic high order internal models can be derived, as it will be shown in the following chapters. Table 2.1 summarizes different internal models commonly used in RC.

2.1.2 The Repetitive Controller

Repetitive controllers are composed by two main elements: the IM, $I(z)$, and the stabilizing controller, $G_x(z)$. The IM is the one in charge of guaranteeing null or small error in steady state, while the stabilizing controller assures closed-loop stability. The standard IM is constructed using (2.4) and setting $W(z) = z^{-N}$ and $\sigma = 1$, resulting in:

$$I_{st}(z) = \frac{H(z)}{z^N - H(z)}. \qquad (2.5)$$

Table 2.1 Some internal models used in repetitive control

IM / Action	Full-harmonic	Odd-harmonic	$6l \pm 1$
RC	$I_{st}(z) = \frac{H(z)}{z^N - H(z)}$	$I_{odd}(z) = \frac{-H(z)}{z^{\frac{N}{2}} + H(z)}$	$I_{6l\pm 1}(z) = \frac{W(z)H(z)}{1 + W(z)H(z)}$
	$W(z) = z^{-N}, \sigma = 1$	$W(z) = z^{-\frac{N}{2}}, \sigma = -1$	$W(z) = z^{-\frac{N}{3}} - z^{-\frac{N}{6}}, \sigma = -1$
HORC	$I_{ho}(z) = \frac{W(z)H(z)}{1 - W(z)H(z)}$	$I_{hodd}(z) = \frac{-W(z)H(z)}{1 + W(z)H(z)}$	
	$W(z) = 1 - (1 - z^{-N})^M, \sigma = 1$	$W(z) = -1 + \left(1 + z^{-\frac{N}{2}}\right)^M, \sigma = -1$	

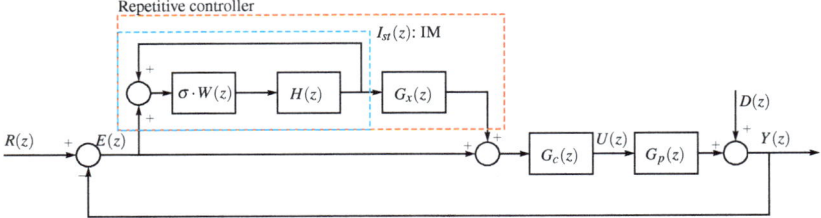

Fig. 2.3 Discrete-time block-diagram of the proposed repetitive transfer function

When $H(z) = 1$, the poles of (2.5) are uniformly distributed over the unit circle[1], $z = \exp(2k\pi j/N)$, providing infinite gain at frequencies $\omega_k = 2k\pi/N$, with $k = 0, 2, ..., N-1$.

Although the IM and the stabilizing controller can be arranged in different ways, most repetitive controllers are usually implemented in a "plug-in" fashion, as depicted in Figure 2.3: the repetitive compensator is used to augment an existing nominal controller, $G_c(z)$. This nominal compensator is designed to stabilize the plant, $G_p(s)$, and tp provide disturbance attenuation across a broad frequency spectrum.

Assume that both T_p and T_s are constant, which makes $N = T_p/T_s$ also constant, and let $G_p(z)$ stand for the corresponding z-transform of $G_p(s)$. Sufficient stability criteria are given in the next Proposition:

Proposition 2.1. *The closed-loop system of Figure 2.3 is stable if the following conditions are fulfilled [3, 10]:*

1. The closed-loop system without the repetitive controller $G_o(z)$ is stable, where

$$G_o(z) = \frac{G_c(z)G_p(z)}{1 + G_c(z)G_p(z)}. \qquad (2.6)$$

2. $\| W(z)H(z)[1 - G_o(z)G_x(z)] \|_\infty < 1$, where $G_x(z)$ is a design filter to be chosen.

The sensitivity function of the closed-loop system depicted in Figure 2.3, using the generic IM (2.4), is:

$$S(z) = \frac{E(z)}{R(z)} = S_o(z)S_{Mod}(z), \qquad (2.7)$$

where

$$S_o(z) = \frac{1}{1 + G_c(z)G_p(z)} \qquad (2.8)$$

stands for the sensitivity function of the system without repetitive controller and $S_{Mod}(z)$ is the modifying sensitivity function

---

[1] Note that, since there is also a pole in $z = 1$, there is infinite gain in dc-frequency, i.e. an integral action.

$$S_{Mod}(z) = \frac{1 - \sigma W(z)H(z)}{1 - \sigma W(z)H(z)(1 - G_x(z)G_o(z))}. \quad (2.9)$$

The closed-loop system poles are the poles of the system without repetitive controller, i.e. poles of $S_o(z)$ and the poles of $S_{Mod}(z)$ defined in equations (2.8) and (2.9), respectively. When $G_x(z) = k_r(G_o(z))^{-1}$, $\sigma = 1$, $W(z) = z^{-N}$ and $H(z) = 1$, the poles of $S_{Mod}(z)$ are:

$$z = \sqrt[N]{|1-k_r|} e^{j(\frac{2k\pi}{N} + \pi \cdot \frac{1 - \text{sgn}(1-k_r)}{2})}, \ k = 0, \ldots, N-1. \quad (2.10)$$

These poles are uniformly distributed over a circle of radius $\sqrt[N]{|1-k_r|}$. To render stability these poles should be within the unit circle, namely $k_r \in (0,2)$. Although the introduction of $H(z) \neq 1$ and of designs that involve nonminimum-phase plants affect the location of the closed-loop poles and also the convergence speed of the system as a function of k_r, this analysis gives a simple and intuitive approximation of the distribution of the poles [17].

Stability Filter $G_x(z)$

Condition 2 of Proposition 2.1 should be fulfilled with an appropriate design of the filter $G_x(z)$. The fundamental issue is to provide enough leading phase to cancel out the phase of $G_o(z)$ [9]. In case of minimum phase systems $G_x(z)$ is implemented as the inverse of the complementary sensitivity function $G_o(z)$.

Gain k_r

The design of the gain k_r involves a trade-off between stability robustness and steady state performance [8]. Furthermore, it has been proved that an appropriate selection of k_r decreases the error caused by non harmonic components [9], and a tuning algorithm has been proposed in [1].

Robustness Filter $H(z)$

Due to model uncertainty and unmodelled dynamics the cancellation performed by the filter $G_x(z)$ is not perfect, specially at high frequencies. Moreover, the allowable model uncertainty is smaller at harmonic frequencies where the IM brings infinite gain, which is more critical at high frequencies [14]. Therefore, the filter $H(z)$ is used to limit the infinite gain mainly at high frequencies, thus improving the stability robustness. In this way, $H(z)$ should provide enough band limitation in accordance with the frequency interval where model uncertainty exists. In general, $H(z)$ is chosen to be a null phase low-pass FIR filter [9, 13, 14]. If conventional low-pass filters are used it will be necessary to compensate for the phase shifting they cause [5, 15].

When limiting the gain by means of the filter $H(z)$ the tracking/rejecting performance decreases mainly at high frequencies and a slight deviation of the generated harmonic frequencies arises due to pole shifting. As happens with other components, the design of $H(z)$ implies a trade-off between stability robustness and performance as well.

Design Procedure

The design of $H(z)$, $G_x(z)$ and $G_c(z)$ should also consider the following issues [3, 10]:

- It is advisable to design the controller $G_c(z)$ with a high enough robustness margin.
- $H(z)$ is designed to have gain close to 1 in the desired bandwidth and attenuate the gain out of it.
- A trivial structure which is often used for $G_x(z)$ in case that $G_o(z)$ is minimum-phase is [12]:

$$G_x(z) = k_r [G_o(z)]^{-1}.$$

Otherwise, alternative techniques should be applied in order to avoid unstable pole-zero cancellations [12] (see Appendix A). Moreover, there is no problem with the improperness of $G_x(z)$ because the IM provides the repetitive controller with a high positive relative degree. Finally, as argued in [8], k_r must be designed looking for a trade-off between robustness against plant uncertainty and transient response.

2.2 Performance under Varying Frequency

The theoretical basics and the design developed above assume that the frequency of the signal to be tracked/rejected is constant and well known. However for most of the practical implementations it is important to analyse the system performance in the face of frequency variations or frequency uncertainty.

The repetitive controller introduced in Section 2.1 contains the ratio $N = T_p/T_s$, which is embedded in the controller implementation. This setting renders a well-known good performance if the reference/disturbance periodic signal has a known constant period T_p. However, the controller performance decays dramatically when a variation of T_p appears. As an example, Figure 2.4 shows the magnitude response of an IM designed for a 50 Hz periodic signal, with the gain for 49 Hz and 51 Hz (and some of their harmonics) highlighted. Note that the gain is large at the fundamental and harmonic frequencies, but it is highly reduced when the frequency is slightly deviated from these frequencies. Consequently, the tracking/rejection capabilities are dramatically reduced. Also, it is worth noticing that the gain reduction is even worse for higher harmonics.

Something similar occurs with the phase lag of the closed-loop control system. Figure 2.5 portrays the closed-loop phase response of a RC implementation using the settings of Section 7.2 but using the full harmonic IM (2.5). It can be seen that while for the nominal frequency the phase is almost zero, this does not happen with the other frequencies. It is worth emphasizing that this would imply a reduction of the reference signal tracking capabilities, which would contribute to the reduction of the system performance. Furthermore, it can be also realized that the phase difference worsens for higher harmonics.

2.2 Performance under Varying Frequency 11

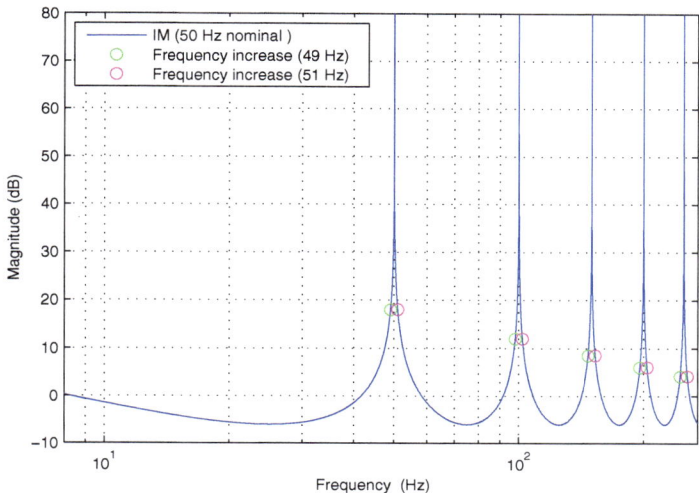

Fig. 2.4 Open-loop IM gain diagram

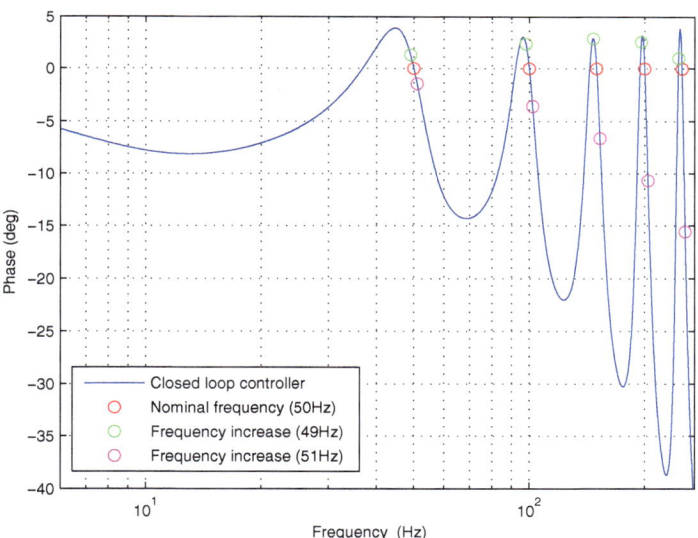

Fig. 2.5 Closed-loop transfer function phase diagram

References

1. Chang, W.S., Suh, I.H., Kim, T.W.: Analysis and design of two types of digital repetitive control systems. Automatica 31(5), 741–746 (1995)
2. Chew, K.K., Tomizuka, M.: Steady-state and stochastic performance of a modified discrete-time prototype repetitive controller. Journal of Dynamic Systems, Measurement, and Control 112, 35–41 (1990)
3. Costa-Castelló, R., Nebot, J., Griñó, R.: Demonstration of the internal model principle by digital repetitive control of an educational laboratory plant. IEEE Transactions on Education 48(1), 73–80 (2005)
4. Escobar, G., Hernandez-Briones, P., Torres-Olguin, R., Valdez, A.: A repetitive-based controller for the compensation of $6l \pm 1$ harmonic components. In: Proceedings of the IEEE International Symposium on Industrial Electronics, pp. 3397–3402 (2007)
5. Escobar, G., Torres-Olguin, R., Valdez, A., Martinez-Montejano, M., Hernandez-Briones, P.: Practical modifications of a repetitive-based controller aimed to compensate 6l+1 harmonics. In: Proceedings of the 11th IEEE International Power Electronics Congress, CIEP 2008, pp. 90–95 (August 2008)
6. Francis, B., Wonham, W.: Internal model principle in control theory. Automatica 12, 457–465 (1976)
7. Griñó, R., Costa-Castelló, R.: Digital repetitive plug-in controller for odd-harmonic periodic references and disturbances. Automatica 19(4), 1060–1068 (2004)
8. Hillerström, G., Lee, R.C.: Trade-offs in repetitive control. Technical Report CUED/F-INFENG/TR 294, University of Cambridge (June 1997)
9. Inoue, T.: Practical repetitive control system design. In: Proceedings of the 29th IEEE Conference on Decision and Control, pp. 1673–1678 (1990)
10. Inoue, T., Nakano, M., Kubo, T., Matsumoto, S., Baba, H.: High accuracy control of a proton synchroton magnet power supply. In: Proceedings of the 8th IFAC World Congress, pp. 216–220 (1981)
11. Leyva-Ramos, J., Escobar, G., Martinez, P., Mattavelli, P.: Analog circuits to implement repetitive controllers for tracking and disturbance rejection of periodic signals. IEEE Transactions on Circuits and Systems II: Express Briefs 52(8), 466–470 (2005)
12. Tomizuka, M., Tsao, T.-C., Chew, K.-K.: Analysis and synthesis of discrete-time repetitive controllers. Journal of Dynamic Systems, Measurement, and Control 111, 353–358 (1989)
13. Tsao, T.-C., Tomisuka, M.: Adaptive and repetitive digital control algorithms for non circular machining. In: Proceedings of the 1988 American Control Conference (1988)
14. Tsao, T.-C., Tomizuka, M.: Robust adaptive and repetitive digital tracking control and application to a hydraulic servo for noncircular machining. Journal of Dynamic Systems, Measurement, and Control 116(1), 24–32 (1994)
15. Weiss, G., Häfele, M.: Repetitive control of MIMO systems using H_∞ design. Automatica 35(7), 1185–1199 (1999)
16. Yamamoto, Y.: Learning control and related problems in infinite-dimensional systems. In: Proceedings of the European Control Conference, pp. 191–222 (1993)
17. Yeol, J.W., Longman, R.W., Ryu, Y.S.: On the settling time in repetitive control systems. In: Proceedings of 17th International Federation of Automatic Control (IFAC) World Congress (July 2008)

Part I

Varying Sampling Approach

3
Stability Analysis Methods

Summary. Repetitive control is a widely used strategy applied in the tracking/rejection of periodic signals, however, the performance of this controller can be seriously affected when the frequency of the reference/disturbance signal varies or is uncertain. One approach that overcomes this problem is the adaptation of the controller sampling period, nevertheless, the system framework changes from a Linear Time Invariant to Linear Time-Varying and the closed-loop stability can be compromised. Indeed, the proposals applying this scheme in repetitive control do not provide formal stability proofs. This work presents two different methodologies aimed at analyse the system stability under these conditions. The first one uses a Linear Matrix Inequality gridding approach which provides necessary conditions for the closed-loop Bounded Input Bounded Output stability of the system. The second one applies robust control techniques in order to analyse the stability and yields sufficient stability conditions. Both methodologies, entails a frequency variation interval for which the system stability can be assured.

3.1 Introduction

3.1.1 State of the Art

The idea of adapting the sampling rate of the repetitive controller according to the reference/disturbance period (see [6, 7]) allows the preservation of a constant value for N and the steady-state performance of the controller. Also, two additional advantages are obtained: the computational cost is kept low and, since the number of samples per period is maintained constant, the quality of the continuous-time signal reconstruction is preserved. However, as mentioned before, this approach implies that the structure of the system changes from LTI to LTV, which may result in closed-loop instability. Indeed, no formal stability proofs are provided in the existing literature for this case. The approach in [7] tries to ensure stability using different controllers according to the value of T_s, but stability is not formally proved. Differently, [6] deals with the specific case of periodic variations of T_p and proves stability after transforming the periodic system into an LTI one by means of lifting techniques. The third proposal [14] considers the adaptation of both N and T_s, and stability is claimed assuming that eventual changes in T_s are small enough.

In a more general context, the stability analysis of sampled-data linear systems with time-varying sampling rates is a challenging problem which may follow several approaches. The first one [11] uses an LMI gridding technique that allows to establish necessary conditions to fulfil a discrete-time sufficient condition. Sufficient stability conditions are reported in [2], where aperiodic sampling is modeled as a piecewise continuous delay in the control input; less conservative conditions are given in [8] after interpreting the obtained stability condition in terms of the small gain theorem, while the use of passivity-type properties yields an additional improvement of the technique [5]. The third approach [9] is based on a hybrid modelling of sampled-data systems and a search of discontinuous Lyapunov functions. In all the above reported works the different stability conditions are established in an LMI format. However, computational issues may arise when trying to solve LMI problems that involve high order systems, and repetitive controllers use to share this feature.

Recently, a different insight has been considered in [3, 13]. The key idea in both contributions, which use a static controller, is to model the non-uniform sampling time effect as a nominal system affected by an additive, norm-bounded time-varying disturbance. Hence, small-gain theorem-based robust control techniques may ensure stability in closed neighborhoods of the nominal sampling period.

3.1.2 Contribution

This Chapter describes tools developed with the aim of analyzing the stability of a system containing a digital repetitive controller working under time-varying sampling period.

Thus, an accurate study of stability margins where reliable performance may be ensured without specifically requiring T_p to be periodic is carried out, thus improving the results in [7] and generalizing the ones in [6]. The stability is studied twofold. The first proposed methodology uses an LMI gridding approach [1, 11] that allows to assess conditions for the BIBO stability of the closed-loop system in a known, bounded interval where the reference/disturbance period is assumed to vary. The second one is based on robust control techniques, which allows to overcome intrinsic lacks of the LMI gridding approach and presents an adaptation and improvement of the approach introduced in [3, 13] for digital repetitive control systems. The key point of the improvement lies on the fact that the proposed technique uses a discrete-time dynamic controller that yields a degenerate disturbance matrix. This allows to carry out an optimal decomposition with regard to the expected stability interval. Furthermore, an additional discussion that may help towards a conservatism reduction of the stability intervals is provided.

3.2 Repetitive Control under Varying Frequency Conditions

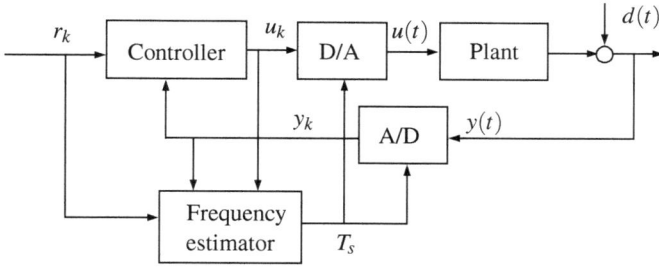

Fig. 3.1 Accommodation of the sampling period T_s to possible variations of T_p

3.1.3 Outline

Section 3.2 describes an adaptation procedure for the sampling period according to the time-variation of the reference/disturbance signal. Section 3.3 analyses the stability of the closed-loop system using the LMI gridding approach and Section 3.4 studies the stability of the system using robust control techniques, while conclusions are presented in Section 3.5.

3.2 Repetitive Control under Varying Frequency Conditions

In this approach the controller sampling time T_s is adapted according to the reference/disturbance period variation T_p, and thus maintaining a constant value for N. Hence, on the one hand, $I(z)$, $G_x(z)$ and $G_c(z)$ are designed and implemented to provide closed-loop stability for a nominal sampling time $T_s = \bar{T}$ (see Chapter 2 for stability criteria and design issues). In this way, their structures remain always invariant, i.e. they undergo no further structural changes. On the other hand, the period of the sampler preceding the plant, $G_p(s)$, is accommodated to the variation of T_p. Therefore, its discrete-time representation corresponds to an LTV system. The accommodation scheme is detailed in Figure 3.1.

Although the proposed technique allows to adapt the system to the specific frequency variation to be tracked/rejected without changing the digital controller, the sampling rate change may compromise closed-loop stability. In what follows, a closed-loop stability analysis of the system under varying sampling rate condition is carried out. For the rest of the chapter it is assumed that the original continuous-time plant is controllable and observable, and also that these properties are not lost with sampling.

Regarding the time-varying nature of the plant sampling period, the stability analysis is carried out in state-space. The final part of the Section is devoted to obtaining the state equations of the system.

Let the discrete-time state-space representations of blocks $I(z)$, $G_x(z)$, $G_c(z)$ and $G_p(z)$ (see Figure 2.3) be denoted by (A_i, B_i, C_i, D_i), with the subindex i replaced by

18   3 Stability Analysis Methods

I, x, c and p, respectively. The closed-loop system state equations are derived under the following assumptions:

- The IM, $I(z)$, of Figure 2.3 is such that $D_I = 0$.
- The continuous-time plant $G_p(s)$ has at least relative degree 1, so $D_p = 0$.
- The representations corresponding to blocks $I(z)$, $G_x(z)$ and $G_c(z)$ are obtained from the nominal sampling time $T_s = \bar{T}$.
- Only the discrete-time plant model matrices A_p, B_p vary according to sampling rate updating i.e. $A_p = A_p(T_s)$, $B_p = B_p(T_s)$ while C_p is kept constant. Hence, assuming that $(A,B,C,0)$ stands for the continuous-time plant state-space representation, i.e. $G_p(s) = C(s\mathbb{I} - A)^{-1} B$, then

$$A_p(T) \triangleq e^{AT}, \quad B_p(T) \triangleq \int_0^T e^{Ar} B\, dr. \quad (3.1)$$

Let the system be sampled at $\{t_0, t_1, \ldots\}$ with $t_0 = 0$, $t_{k+1} > t_k$, the sampling periods being $T_k = t_{k+1} - t_k$. Let also $x_k \triangleq x(t_k)$, $r_k \triangleq r(t_k)$, $y_k \triangleq y(t_k)$. The state equations are given by:

$$x_{k+1} = \Phi(T_k) x_k + \Pi(T_k) r_k, \quad y_k = \Upsilon x_k, \quad (3.2)$$

where

$$\Phi(T) \triangleq \begin{pmatrix} K & L \\ B_p(T)M & A_p(T) + B_p(T)Q \end{pmatrix}, \quad (3.3)$$

with

$$K \triangleq \begin{pmatrix} A_I & 0 & 0 \\ B_x C_I & A_x & 0 \\ B_c D_x C_I & B_c C_x & A_c \end{pmatrix}, \quad L \triangleq \begin{pmatrix} -B_I C_p \\ 0 \\ -B_c C_p \end{pmatrix},$$

$$M \triangleq \begin{pmatrix} D_c D_x C_I & D_c C_x & C_c \end{pmatrix}, \quad Q \triangleq -D_c C_p$$

and

$$\Pi(T) \triangleq \begin{pmatrix} B_I^\top & 0 & B_c^\top & (B_p(T)C_c)^\top \end{pmatrix}^\top, \quad \Upsilon \triangleq \begin{pmatrix} 0 & 0 & 0 & C_p \end{pmatrix}.$$

Remark 3.1. When T_k remains constant, i.e. $T_k = \bar{T}$, $\forall k \geq 0$, (3.2) becomes a discrete-time LTI system with z-transfer function

$$G(z) = \Upsilon \left[z\mathbb{I} - \Phi(\bar{T}) \right]^{-1} \Pi(\bar{T}).$$

In an aperiodic sampling time framework, $\Phi(T_k)$ and $\Pi(T_k)$ vary with k, and the z-transform representation is no longer valid.

In RC systems, the IM is designed to ensure a null error in the steady state provided that closed-loop stability is guaranteed. Hence, assuming that $I(z)$, $G_x(z)$ and $G_x(z)$ are constructed to provide stability for a nominal sampling time $T_s = \bar{T}$, the aim is to prove both internal and BIBO stability for the non-uniformly sampled system (3.2).

Throughout this book $\|\cdot\|$ denotes the 2-norm of a matrix, i.e. the matrix norm induced by the euclidean vector norm. Hence, for any real matrix

R, $\|R\| = \left[\rho(R^\top R)\right]^{1/2}$, with $\rho(\cdot)$ standing for the spectral radius. Moreover, recall that given a discrete-time LTI system with constant sampling period T_s and transfer function matrix $G(z)$, its H_∞-norm is defined by

$$\|G(z)\|_\infty = \max\left\{\|G\left(e^{j\omega T_s}\right)\|,\ \forall \omega \in \mathbb{R}\right\}.$$

Proposition 3.1. *Let the sampling period, T_k, take values in a compact subset $\mathscr{T} \subset \mathbb{R}^+$. Then, the uniform exponential stability of*

$$x_{k+1} = \Phi(T_k)x_k \qquad (3.4)$$

implies the uniform BIBO stability of (3.2).

Proof. According to Lemma 27.4 in [10], the result follows if $\Pi(T_k)$ and Υ are uniformly bounded matrices, $\forall k \geq 0$, and this is indeed true: $\Pi(T_k)$ depends continuously on T_k, which belongs to a compact set $\mathscr{T}$, while Υ is a constant matrix because so is C_p. □

Proposition 3.2 ([10]). *Let the sampling period, T_k, take values in a compact subset $\mathscr{T} \subset \mathbb{R}^+$. If there exists a matrix P such that*

$$\Phi(T_k)^\top P \Phi(T_k) - P < 0,\ \forall T_k \in \mathscr{T},\ s.t.\ P = P^\top > 0, \qquad (3.5)$$

then (3.4) is uniformly exponentially stable.

3.3 LMI Gridding Approach

It is immediate that relation (3.5) in Proposition 3.2 yields an infinite set of LMIs. The gridding approach introduced in [1, 11] allows a simple stability analysis that may be performed in two stages, if necessary.

In a first stage, advantage is taken from the fact that (3.2) is stable by construction for $T_k = \bar{T}$, $\forall k$.

Proposition 3.3. *Assume that the stability conditions of Proposition 2.1 are satisfied for a nominal sampling period $\bar{T} \in \mathscr{T}$. Then,*

1. *The zero state of the LTI system (3.4) with $T_k = \bar{T}$, $\forall k$, is uniformly exponentially stable.*
2. *The LMI problem*

$$L_{\bar{T}}(P) \leq -\alpha \mathbb{I},\ s.t.\ P = P^\top > 0, \qquad (3.6)$$

   with $L_{\bar{T}}(P)$ constructed from (3.5) and $\alpha \in \mathbb{R}^+$, is feasible.
3. *Let $P = P_N$ be a solution of the LMI problem (3.6) for a fixed $\alpha \in \mathbb{R}^+$. Then, there exists an open neighborhood of $\bar{T}$, say $\mathscr{I}_N$, such that (3.4) is BIBO stable for all T_k in $\mathscr{I}_N$.*

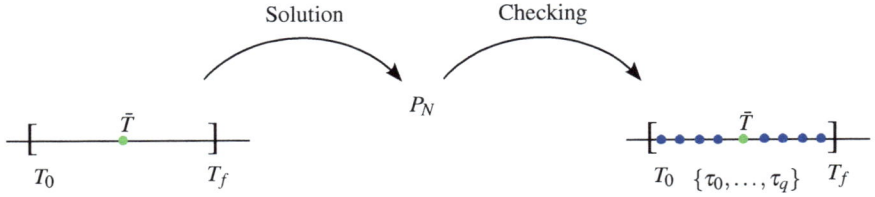

Fig. 3.2 Graphical scheme of the LMI gridding-based stability analysis in $\mathscr{T} = [T_0, T_f]$ from the solution of (3.6)

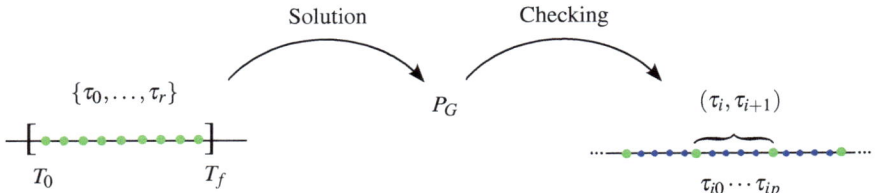

Fig. 3.3 Graphical scheme of the LMI gridding-based stability analysis in $\mathscr{T} = [T_0, T_f]$ from the solution of (3.7)

Proof. It follows from the stability hypothesis that all the eigenvalues of $\Phi(\bar{T})$ are within the unit circle, which yields item 1 [10]. Item 2 stems from the fact that the sufficient condition for uniform exponential stability established in Proposition 3.2 is also necessary for a discrete-time LTI system. Finally, item 3 follows from Propositions 3.1 and 3.2 taking into account the continuity of the matrix elements of $\Phi(T_k)$ with respect to T_k. □

Assume that we are interested in studying the stability of (3.4) for all sampling periods $T_k \in \mathscr{T}$. Let $P = P_N$ be a feasible solution of the LMI problem (3.6) for a certain $\alpha \in \mathbb{R}^+$, its existence being guaranteed by Proposition 3.3. Let also $\{\tau_0, \ldots, \tau_q\}$, with $\tau_{i+1} > \tau_i$, be a sufficiently fine grid of $\mathscr{T}$. Then, one has to check that $L_{\tau_i}(P_N) < 0$, $\forall i = 0, \ldots, q$: see Figure 3.2.

In case that there exists at least a single τ_i such that $L_{\tau_i}(P_N) \geq 0$, the gridding procedure proposed in [11] may be carried out as follows. Let $\{\tau_0, \ldots, \tau_r\}$, be a sorted set of sampling period candidates suitably distributed in $\mathscr{T}$. Then, one may solve the finite set of LMIs:

$$L_{\tau_i}(P) \leq -\alpha \mathbb{I}, \quad i = 0, \ldots, r, \quad \text{s.t.} \quad P = P^\top > 0, \tag{3.7}$$

for a fixed $\alpha \in \mathbb{R}^+$. If the problem is feasible and a solution, $P = P_G$, is found out, the negative-definite character of $L_{T_k}(P_G)$ has to be checked for intermediate values of T_k in each open subinterval (τ_i, τ_{i+1}), i.e. for all $\{\tau_{ij}; i = 0, \ldots, r-1, \; j = 1, \ldots, p\}$, with $\tau_i < \tau_{i1} < \cdots < \tau_{ip} < \tau_{i+1}$, for all $i = 0, \ldots, r-1$; see Figure 3.3. In case that this is not accomplished, (3.7) has to be solved again for a finer grid of $\mathscr{T}$. Otherwise, the procedure should be run for a new interval $\mathscr{T}' \subset \mathscr{T}$.

Remark 3.2. Notice that:

(i) Comparatively, in the first approach one may expect a shorter stability radius.

(ii) An important drawback of both analysis is associated to the fact that stability is rigorously guaranteed just in open neighbourhoods of each sampling period τ_i or τ_{ij} where either $L_{\tau_i}(P_N) < 0$ or $L_{\tau_{ij}}(P_G) < 0$ are satisfied, respectively, thus indicating a partial fulfilment of the sufficient stability condition. However, the distance between these intervals may be reduced at will by increasing the number of grid points. Furthermore, it is important to note that in a real application the sampling time variation is defined by a finite precision digital clock which yields a variation interval with a finite number of points. Therefore, including these points to check the Lyapunov condition may guarantee the stability of the digital implementation in a defined interval.

(iii) RC systems use to be high order systems, and this fact may yield computational issues when solving LMIs. Hence, in the next subsection an alternative tool for stability analysis is introduced.

(iv) A faulty estimation of T_p and/or an error in the implementation of T_s may yield important performance degradation [12]. However, closed-loop BIBO stability is not threatened unless the plant is actually sampled with T_s values lying outside the region $\mathscr{T}$ where stability is guaranteed.

3.4 Robust Analysis

The stability analysis follows the approach proposed in [4, 13], where the non-uniform sampling is regarded as a nominal sampling period affected by an additive disturbance. Thus, the problem is faced from a robustness analysis viewpoint and is solved by means of the small-gain theorem. As in this case. The RC system is designed to provide closed-loop stability for a nominal sampling, the actual problem is to quantify the amount of disturbance due to aperiodic sampling that the system can accommodate while preserving stability.

Proposition 3.4. *Let $\bar{T}$ be a nominal sampling period and define $\theta_k = T_k - \bar{T}$. Then, the matrix $\Phi(T_k)$ may be written as*

$$\Phi(T_k) = \Phi(\bar{T}) + \Gamma \tilde{\Delta}(\theta_k) \Psi(\bar{T}), \tag{3.8}$$

where

$$\tilde{\Delta}(\theta_k) \triangleq \begin{pmatrix} 0 & 0 \\ 0 & \Delta(\theta_k) \end{pmatrix}, \quad \Delta(\theta_k) \triangleq \int_0^{\theta_k} e^{Ar} dr, \tag{3.9}$$

$$\Psi(T_k) \triangleq \begin{pmatrix} 0 & 0 \\ A_p(T_k)A & A_p(T_k)B \end{pmatrix} \begin{pmatrix} 0 & \mathbb{I} \\ M & Q \end{pmatrix} \tag{3.10}$$

and $\Gamma \in \Xi$, with

$$\Xi := \{\Gamma; \Gamma\tilde{\Delta}(\theta) = \tilde{\Delta}(\theta)\} = \left\{ \begin{pmatrix} \Gamma_1 & 0 \\ \Gamma_2 & \mathbb{I} \end{pmatrix} \right\}. \tag{3.11}$$

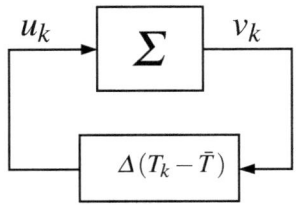

Fig. 3.4 The resulting feedback system

Proof. Recalling (3.1) and using

$$e^{A\theta} = \mathbb{I} + \int_0^\theta e^{Ar} A \, dr,$$

one has that

$$A_p(T_k) = e^{A(\theta_k + \bar{T})} = \left(\mathbb{I} + \int_0^{\theta_k} e^{Ar} A \, dr\right) A_p(\bar{T}) = A_p(\bar{T}) + \Delta(\theta_k) A_p(\bar{T}) A; \quad (3.12)$$

on the other hand,

$$B_p(T_k) = \int_0^{T_k} e^{Ar} B \, dr = \int_0^{\bar{T}} e^{Ar} B \, dr + \int_{\bar{T}}^{T_k} e^{Ar} B \, dr = B_p(\bar{T}) + \int_0^{\theta_k} e^{A(s+\bar{T})} B \, ds =$$
$$= B_p(\bar{T}) + \Delta(\theta_k) A_p(\bar{T}) B. \quad (3.13)$$

The result follows immediately taking (3.12) and (3.13) to (3.3). □

Notice that, using Proposition 3.4, (3.4) becomes

$$x_{k+1} = \left[\Phi(\bar{T}) + \Gamma \tilde{\Delta}(\theta_k) \Psi(\bar{T})\right] x_k, \quad (3.14)$$

which allows the following interpretation [4]: (3.14) can be regarded as the LTI system

$$\Sigma := \begin{cases} x_{k+1} = \Phi(\bar{T}) x_k + \Gamma u_k, \\ v_k = \Psi(\bar{T}) x_k, \end{cases} \quad (3.15)$$

$G_{\bar{T}}(z) = \Psi(\bar{T}) \left[z\mathbb{I} - \Phi(\bar{T})\right]^{-1} \Gamma$ being its associated discrete-time transfer function, receiving the time-varying output feedback control action $u_k = \tilde{\Delta}(\theta_k) v_k$ as depicted in Figure 3.4.

Theorem 3.1. *Assume that $\bar{T}$ is a nominal sampling period. Let*

$$\gamma_{\bar{T}} = (1+\varepsilon)\|G_{\bar{T}}(z)\|_\infty, \quad \varepsilon > 0, \quad (3.16)$$

be an upper bound of the H_∞-norm of system Σ (3.15), and let also $\mathscr{T} \subset \mathbb{R}^+$ be compact. If

$$\gamma_{\bar{T}} \|\Delta(T_k - \bar{T})\| \leq 1, \quad \forall T_k \in \mathscr{T}, \quad (3.17)$$

then system (3.2) is uniformly BIBO stable in $\mathscr{T}$.

Proof. As $\bar{T}$ is a nominal sampling period, $\rho\left[\Phi(\bar{T})\right] < 1$, and the specific block-diagonal structure of $\tilde{\Delta}(\theta_k)$ (see (3.9)) indicates that $\|\Delta(T_k - \bar{T})\| = \|\Delta(\theta_k)\| = \|\tilde{\Delta}(\theta_k)\|$. According to Lemma 2 in [4], the stated hypotheses are sufficient for the LMI problem (3.5) to be feasible $\forall T_k \in \mathscr{T}$. Then, Proposition 3.2 entails the uniform exponential stability of (3.4) and, therefore, Proposition 3.1 yields the uniform BIBO stability of (3.2). □

Remark 3.3. In sight of Theorem 3.1, the obtention of wider stability intervals may depend on the following issues:

1. Minimization of $\gamma_{\bar{T}}$.
   It may be achieved through the selection a matrix $\Gamma \in \Xi$ that minimizes $\|G_{\bar{T}}(z)\|_\infty = \|G_{\bar{T}}(z,\Gamma)\|_\infty$. This problem is faced in Proposition 3.5 below.
2. Reduction of conservatism in the calculation of the interval $\mathscr{T}$ where $\|\Delta(T - \bar{T})\| \leq \gamma_{\bar{T}}^{-1}$ is ensured.
   Recalling (3.9), it is immediate that the repetitive control system has no influence in this point; indeed, it depends solely on the plant.

Proposition 3.5. *A solution of the problem*

$$\min\left\{ \|\Psi(T)\left[z\mathbb{I} - \Phi(T)\right]^{-1}\Gamma\|_\infty, \ \Gamma \in \Xi \right\},$$

is given by

$$\Gamma = \begin{pmatrix} 0 & 0 \\ 0 & \mathbb{I} \end{pmatrix}. \qquad (3.18)$$

Proof. Assuming that $\Gamma \in \Xi$ (see (3.11)) and writing $\Psi(T)$, $\left[z\mathbb{I} - \Phi(T)\right]^{-1}$, in block form as

$$\Psi(T) = \begin{pmatrix} 0 & 0 \\ \Psi_1 & \Psi_2 \end{pmatrix}, \quad \left[z\mathbb{I} - \Phi(T)\right]^{-1} = \begin{pmatrix} F_{11} & F_{12} \\ F_{21} & F_{22} \end{pmatrix},$$

where $\Psi_i = \Psi_i(T)$ and $F_{ij} = F_{ij}(z,T)$, it results that

$$G_T(z,\Gamma) = \begin{pmatrix} 0 & 0 \\ G_{T1}(z,\Gamma_1,\Gamma_2) & G_{T2}(z) \end{pmatrix},$$

with

$$G_{T1} = (\Psi_1 F_{11} + \Psi_2 F_{21})\Gamma_1 + (\Psi_1 F_{12} + \Psi_2 F_{22})\Gamma_2,$$
$$G_{T2} = \Psi_1 F_{12} + \Psi_2 F_{22}.$$

Hence, the definition of the H_∞-norm and the fact that matrix expansion does not decrease norms yield straightforward that $\|G_T(z)\|_\infty \geq \|G_{T2}(z)\|_\infty$. □

Remark 3.4. In [3, 13], the decomposition (3.8) uses $\Gamma = \mathbb{I}$. This is indeed the only possibility when $\tilde{\Delta}(\theta)$ is a non-singular matrix, but it becomes a non optimal choice when this is not the case, as shown in Proposition 3.5. Hence, the proposed procedure is generalizable to any system matching (3.8) and possessing a degenerate disturbance matrix (3.9). This encompasses the class of systems that exhibit the block structure portrayed in Figure 2.3 but operate with a generic discrete-time dynamic controller.

Finally, the procedure to calculate the sampling period variation intervals based on Theorem 3.1 using numeric calculation and norm bounds is described in Appendix B.

3.5 Conclusions

This Chapter analyzed the stability of digital RC systems subjected to references/disturbances with time-varying period T_p. The approach proposed a real-time adaptation of the sampling time of the system, T_s, in order to keep $N = T_p/T_s$ at a constant value. The stability issue was analyzed by means of an LMI gridding method and also using robust control techniques.

The LMI gridding approach is rather simple, but it just provides necessary conditions for a sufficient stability condition and becomes unsuitable for high order systems and/or large analysis intervals due to computational limitations. On the contrary, the robust control approach provides better quality results, i.e. sufficient conditions and wider stability margins, for the proposed adaptation structure.

Finally, it is worth mentioning that here the repetitive controller and the frequency observer that allows the real-time updating of T_s are completely decoupled from a stability analysis viewpoint, which allows an independent design.

References

1. Apkarian, P., Adams, R.: Advanced gain-scheduling techniques for uncertain systems. IEEE Transactions on Control Systems Technology 6(1), 21–32 (1998)
2. Fridman, E., Seuret, A., Richard, J.: Robust sampled-data stabilization of linear systems: An input delay approach. Automatica 40(8), 1441–1446 (2004)
3. Fujioka, H.: Stability analysis for a class of networked-embedded control systems: A discrete-time approach. In: Proceedings of the American Control Conference, pp. 4997–5002 (2008)
4. Fujioka, H.: A discrete-time approach to stability analysis of systems with aperiodic sample-and-hold devices. IEEE Transactions Automat. Control 54(10), 2440–2445 (2009)
5. Fujioka, H.: Stability analysis of systems with aperiodic sample-and-hold devices. Automatica 45(3), 771–775 (2009)
6. Hanson, R.D., Tsao, T.-C.: Periodic sampling interval repetitive control and its application to variable spindle speed noncircular turning process. Journal of Dynamic Systems, Measurement, and Control 122(3), 560–566 (2000)
7. Hillerström, G.: On Repetitive Control. PhD thesis, Lulea University of Technology (November 1994)
8. Mirkin, L.: Some remarks on the use of time-varying delay to model sample-and-hold circuits. IEEE Transactions Autom. Control 52(6), 1109–1112 (2007)
9. Naghshtabrizi, P., Hespanha, J., Teel, A.: On the robust stability and stabilization of sampled-data systems: A hybrid system approach. In: Proceedings of the 45th IEEE Confecence on Decision and Control, pp. 4873–4878 (2006)
10. Rugh, W.: Linear system theory, 2nd edn. Prentice-Hall, Inc., Upper Saddle River (1996)

11. Sala, A.: Computer control under time-varying sampling period: An LMI gridding approach. Automatica 41(12), 2077–2082 (2005)
12. Steinbuch, M.: Repetitive control for systems with uncertain period-time. Automatica 38(12), 2103–2109 (2002)
13. Suh, Y.: Stability and stabilization of nonuniform sampling systems. Automatica 44(12), 3222–3226 (2008)
14. Tsao, T.-C., Qian, Y.-X., Nemani, M.: Repetitive control for asymptotic tracking of periodic signals with an unknown period. Journal of Dynamic Systems, Measurement, and Control 122(2), 364–369 (2000)

4
Design Methods

Summary. The tracking/rejection of periodic signals constitutes a wide field of research in the control theory and applications area and Repetitive Control has proven to be an efficient way to face this topic; however, in some applications the period of the signal to be tracked/rejected change in time which cause and important performance degradation in the standard repetitive controller. One technique that can be used to overcome this problem is the adaptation of the controller sampling period, nevertheless this involves an Linear Time Varying scenario where complexity of the analysis and design of the system is highly increased. The previous Chapter developed a system stability analysis trough Linear Matrix Inequality gridding approach and robust control based techniques. Although several approaches exist for the stability analysis of general time-varying sampling period controllers few of them allow an integrated controller design which assures closed-loop stability under such conditions. In this Chapter two design methodologies are presented which assure the system stability of the repetitive control system working under varying sampling period for a given frequency variation interval: a μ-synthesis technique and a pre-compensation strategy.

4.1 Introduction

The previous chapter analyses the stability of RC systems in which the sampling period is adjusted as a way to preserve the performance under varying frequency conditions. As a result, a frequency variation interval can be found where the stability of the system can be guaranteed. As an opposite problem, this chapter studies methodologies by which a controller can be designed assuring stability and steady-state performance of the system for a given frequency variation range. Two different approaches are presented: a μ-synthesis technique and a pre-compensation strategy.

4.1.1 State of the Art

This section propounds an adaptive variation of the sampling period T_s of the RC system [2, 4, 5, 11] aiming at maintaining a constant value for N. This means that the number of samples per period is kept constant, which preserves the continuous-time signal reconstruction quality. However, it implies changes in the system dynamics and, specifically, in the plant model. Therefore, it is important to check that these changes do not threat stability. Although several approaches exist for the stability analysis of time-varying sampling period controllers [3, 8], few of them allow an integrated controller design which assures closed-loop stability under such conditions. An H_∞ Linear Parameter Varying (LPV) design is proposed in [6, 7]: the system sampling period dependency is approximated and later a sampling period dependent controller is obtained through LMI optimization. However, the application of these approaches in the RC field is constrained by the high order usually exhibited by repetitive controllers and the limitations of current LMI solvers.

4.1.2 Contribution

The main contribution of this section involves two designs:

First, the re-design of an original repetitive controller in such a way that the effect of time-varying sampling is modeled as an structured uncertainty. Under this fact, BIBO stability can be assured when the uncertainty belongs to a known bounded set using small-gain theorem-based standard robust control methodologies. Also, steady-state performance is guaranteed when T_s remains constant for sufficiently large time intervals.

Alternatively, a pre-compensation scheme that forces the inner plant to remain invariant despite sampling rate changes is proposed. As a consequence, standard LTI methods can be used in the control design and the stability analysis. Thus, this adaptation strategy includes an additional compensator placed between the controller and the plant that, under the assumption of internal stability, annihilates the effect of the time-varying period and forces the closed-loop behavior to match that of a pre-selected nominal sampling period. Since this strategy involves a plant inversion it is only applicable in case of stable minimum phase plants.

4.1.3 Outline

This Chapter is organized as follows. Section 4.2 presents the design that uses robust control techniques to ensure stability of the RC system for a pre-defined frequency variation interval. Section 4.3 describes a pre-compensation based design to accomodate the sampling time variation effects and, finally, conclusions are outlined in Section 4.4.

4.2 Robust Stability Design

In this Section, a RC design method is described in which the controller is obtained to assure stability in a predefined sampling period variation interval $\mathscr{T}$. Thus, it is proposed to re-design the original repetitive controller in order to achieve that

$$\|G_{\tilde{T}_s}(z)\|_\infty < (1+\varepsilon)^{-1}\|\Delta\,(T_s-\bar{T})\|_\infty^{-1}$$

(recall that $G_{\tilde{T}_s}(z)$ is the transfer function of the system Σ that corresponds to the system rearrangement in Figure 3.4).

In order to compute $\|\Delta\,(T_s-\bar{T})\|_\infty$, it is assumed that BIBO stability is to be preserved $\forall\,T_s \in \mathcal{T}$, $\mathcal{T}$ being a closed interval[1]. Then, a Schur decomposition-derived bound [10] for the matrix exponential (see (3.9)) may be used to compute $\|\Delta\,(T_s-\bar{T})\|_\infty$ in $\mathcal{T}$.

The repetitive controller elements $G_c(z)$ and $G_x(z)$, i.e. the nominal controller and the stabilizing controller, will be re-designed using robust control tools so as to achieve BIBO stability of the closed-loop system in the overall expected interval of variation of the sampling period, $\mathcal{T}$. However, as $G_x(z)$ depends on $G_c(z)$, this results in a coupled problem which may not be easily solvable. Hence, the procedure proposed here is to perform a two-stage re-desing: firstly of $G_c(z)$ -if necessary-, and then of $G_x(z)$.

Remark 4.1. The use of the z-transform in an LTV framework is not formally correct. However, in this chapter the notation is preserved for simplicity and compactness. Therefore, z^{-1} should be understood as a one sample period delay, but with a possible change of sampling interval length from sample to sample.

The possible re-design of the nominal controller, $G_c(z)$, aims at making easier the solution of the optimization problem that will be later posed to obtain $G_x(z)$. Recalling Section 2.1, the first condition required for the stability of the inner control block is that the closed-loop without the repetitive controller is stable, i.e. that $G_o(z)$ defined in (2.6) is stable. Therefore, one should check that the proposed nominal controller $G_c(z)$ yields stability of $G_o(z)$ not only for the nominal sampling period, $\bar{T}$, but also for all $T_s \in \mathcal{T}$. This can be worked out using the approach in Subsection 3.4, that is, modelling the plant variation due to sampling period changes as an uncertainty and assessing the fulfilment of a (3.17)-like inequality. If this were not the case, $G_c(z)$ should be re-designed by means of a traditional H_∞ problem formulation.

The second stage considers the re-design of $G_x(z)$, aiming at providing stability when including the IM in the closed-loop system using the plug-in architecture introduced in Section 2.1 and taking into account the uncertainty due to sampling period variation. Notice that this could be tackled through an H_∞ problem, but the inclusion of the IM might yield a very high order controller. This problem will be overcome using the approach introduced in [12]. The important delay existing in the IM will be treated as an external uncertainty. Although this might introduce a certain degree of conservativeness, it will also generate a lower order controller.

In order apply this scheme to the complete control-loop, this will be opened in the IM delay and the uncertainty due to sampling period variation, as shown in Figure 4.1. This representation contains three pairs of input-output variables: (u_1, y_1) and

---
[1] From an assumed frequency variation in the interval $[f_{min}, f_{max}]$ Hz, recalling the relation $T_p = N \cdot T_s$, one finds out that $\mathcal{T} = \left[(f_{max}N)^{-1}, (f_{min}N)^{-1}\right]$.

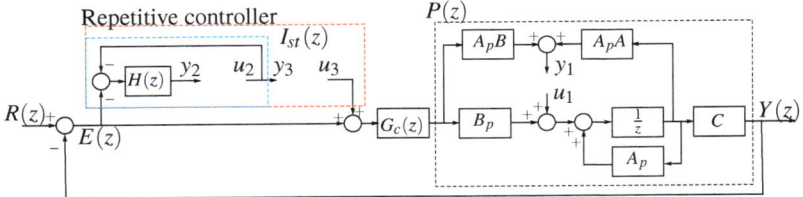

Fig. 4.1 Adaptation of the repetitive control loop to the H_∞ formulation

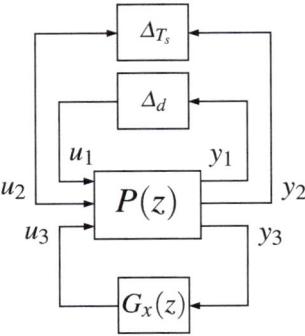

Fig. 4.2 Equivalent scheme

(u_2, y_2), which define the system performance, and (u_3, y_3), which is used to apply the control law. Moreover, for closed-loop stability to be assured it is sufficient that the H_∞ norm from u_1 to y_1 is less than the uncertainty bound in the target sampling interval, $\mathcal{T}$, and also that the H_∞ norm from u_2 to y_2 is less than one. Reducing these two norms will generate a faster and more robust system in the face of sampling period variations.

Figure 4.2 shows the LFT formulation of this problem, the generalized plant being

$$\begin{bmatrix} Y_1(z) \\ Y_2(z) \\ Y_3(z) \end{bmatrix} = P(z) \begin{bmatrix} U_1(z) \\ U_2(z) \\ U_3(z) \end{bmatrix}, \qquad (4.1)$$

with

$$P(z) = \begin{bmatrix} A_pAM(z) - (A_pB + A_pAM(z)B_p)G_1(z)CM(z) & 0 & (A_pB + A_pAM(z)B_p)G_1(z) \\ H(z)CM(z)(G_o(z) - 1) & H(z) & -H(z)G_o(z) \\ 0 & 1 & 0 \end{bmatrix} \qquad (4.2)$$

where $M(z) = (z\mathbb{I} - A_p)^{-1}$, $G_o(z) = G_p(z)G_c(z)/(1 + G_p(z)G_c(z))$ and $G_1(z) = G_c(z)/(1 + G_p(z)G_c(z))$, considering that the two last transfer functions SISO systems. Although this could be addressed using a mixed H_∞ formulation, a μ-synthesis approach can take advantage of the problem structure [9].

Finally, it is worth pointing out that, according to the IMP, steady-state performance will be guaranteed when T_p (and, consequently, T_s) remains constant for large enough time intervals.

4.3 Plant Pre-compensation

This Section presents a design strategy for a digital repetitive controller operating under time-varying sampling period that neutralizes the structural changes caused by the adaptation of the sampling rate. This is achieved with the introduction of a compensator that annihilates the effect of the time-varying sampling and forces the closed-loop behavior to match that of the system operating under a constant and a priory selected nominal sampling period. Hence, once the internal stability of the compensator-plant subsystem is ensured, both time and frequency responses of the overall closed-loop system can be featured using standard LTI tools.

Let us now detail the key issues that support the proposed adaptive compensation scheme:

1. The repetitive controller is designed and implemented to provide closed-loop stability for an a priori selected nominal sampling period $T_s = \bar{T}$, to the nominal LTI plant

$$G_p(z,\bar{T}) \triangleq \frac{Num(z,\bar{T})}{Den(z,\bar{T})}, \qquad (4.3)$$

   in accordance with Proposition 2.1 and the design procedure introduced in Section 2.1.2. Hence, $I_{st}(z)$, $G_x(z)$ and $G_c(z)$ are kept invariant.

2. Aiming at maintaining a constant value N for the ratio T_p/T_s, which preserves the reference/disturbance signal reconstruction quality, the controller sampling rate T_s is accommodated to the time variation of the reference/disturbance period T_p, i.e. $T_s = T_p/N$. Therefore, the discrete-time representation of the plant $G_p(s)$ is that of an LTV system

$$G_p(z,T_s) = \frac{Num(z,T_s)}{Den(z,T_s)}. \qquad (4.4)$$

3. The structural changes caused by the variation of T_s are neutralized with the additional compensator

$$C(z,T_s) = G_p(z,\bar{T})G_p^{-1}(z,T_s), \qquad (4.5)$$

   which pre-multiplies the LTV plant $G_p(z,T_s)$. Thence, under the assumption of internal stability for the compensator-plant system, its behavior is that of the nominal LTI system $G_p(z,\bar{T})$. Figure 4.3 portrays the overall system, while Figure 4.4 details the compensator-plant subsystem.

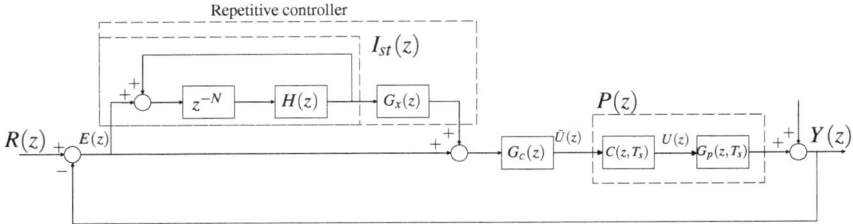

Fig. 4.3 Discrete-time block-diagram of the closed-loop system with the proposed repetitive controller structure

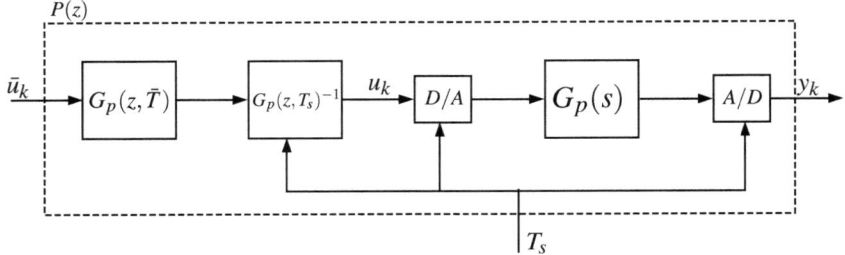

Fig. 4.4 Detail of the compensator-plant system

Remark 4.2. The compensation strategy yields

$$P(z) \triangleq C(z,T_s)G_p(z,T_s) = G_p(z,\bar{T}), \tag{4.6}$$

i.e. a time-invariant block representing the nominal plant for which the repetitive controller provides closed-loop stability. Furthermore, when T_s remains constant at the nominal sampling time, $\bar{T}$, the compensator is such that $C(z,\bar{T}) = 1$.

Notice also that the compensator has no causality problems. Indeed, the use of (4.3) and (4.4) in (4.5) allows to write

$$C(z,T_s) = \frac{Num(z,\bar{T})}{Den(z,\bar{T})} \cdot \frac{Den(z,T_s)}{Num(z,T_s)}. \tag{4.7}$$

Therefore, any possible improperness of the second factor will be removed by the properness of the first factor.

One last consequence is that the response of $P(z)$ at the sampling instants $\{t_k\}_{k\geq 0}$ for a fixed sampling period is the same despite the sampling rate value. This corresponds to a time-scaling effect as a result of forcing the system invariance in the discrete-time space.

The stability of the proposed control scheme is ensured by the following result:

Proposition 4.1. *Assume that the repetitive controller elements $I_{st}(z)$, $G_x(z)$, $G_c(z)$ are designed according to Proposition 2.1 and the design procedure introduced in section 2.1.2 so as to provide of closed-loop stability for the LTI system $G_p(z,\bar{T})$. If the subsystem $P(z)$ pictured in Figure 4.4 is internally stable for all $T_s \in \mathscr{T} \subset \mathbb{R}^+$, then the overall closed-loop system depicted in Figure 4.3 is stable for all $T_s \in \mathscr{T}$.*

Proof. Recalling (4.6), the internal stability of $P(z)$ for all sampling periods $T_s \in \mathscr{T}$ guarantees that compensation is effectively carried out in $\mathscr{T}$. Hence, the overall subsystem $P(z)$ is ensured to behave as the LTI system $G_p(z,\bar{T})$ and, as the repetitive controller is assumed to provide of closed-loop stability to this block, the result follows. □

Remark 4.3. (i) The internal stability of the compensator-plant subsystem $P(z)$ can be checked using an LMI gridding approach [1, 8]. Notice also that a necessary condition for the internal stability of $P(z)$ is that $G_p(z,T_s)$ be minimum phase for all $T_s \in \mathscr{T}$. (ii) In many practical applications the plant $G_p(z)$ admits a first order model description with relative degree 1 which ensures internal stability for $P(z)$ whenever $G_p(z)$ is stable. (iii) The proposed procedure guarantees the expected behaviour at the sampling time instants. However, for extremely large values of $|\theta_k| = |T_k - \bar{T}|$, undesirable intersampling oscillations might arise.

4.4 Conclusions

The adaptation of the sampling period of a repetitive controller in case of reference/disturbance periodic signals with time-varying period allows to preserve its tracking/rejection performance. However, this strategy may negatively affect the closed-loop stability. In this direction, two RC designs that assure the stability of the system working with sampling period adaptation were presented.

First, a controller was obtained through a small-gain theorem-based robust control design technique that assures BIBO stability in the required sampling period interval, with no restrictions on its rate of change. This interval is set from the expected interval of variation, which is assumed to be known.

Second, the pre-compensation design strategy presented in this work propounds to annihilate the structural changes due to the sampling period adjustment by means of an additional compensator which makes the system LTI from an input-output point of view. This allows the characterization of time and frequency responses of the overall closed-loop system using standard LTI tools, provided that internal stability of the compensator-plant subsystem is guaranteed. Furthermore, with this approach, the repetitive controller design defined for a constant sampling period remains valid also in case of aperiodic sampling. The proposed method ensures closed-loop stability independently of the frequency observer dynamics; thence, a decoupled independent design is feasible.

In case of perfect estimation of the signal frequency, the model invariance obtained with the pre-compensation scheme makes possible to preserve performance during frequency transients. Furthermore, the proposed scheme allows transient analysis in a LTI framework.

The pre-compensation scheme is subject to the following limitations:

(i) As an inversion is carried out, it cannot be used with nonminimum phase plants.
(ii) Internal stability is required for the compensator-plant subsystem, which in general needs a non-trivial stability analysis regarding its LTV nature. As the subsystem does not usually exhibit high order, the LMI gridding procedure described in Section 3.3 is suitable for the stability analysis.

References

1. Apkarian, P., Adams, R.: Advanced gain-scheduling techniques for uncertain systems. IEEE Transactions on Control Systems Technology 6(1), 21–32 (1998)
2. Costa-Castelló, R., Malo, S., Griñó, R.: High performance repetitive control of an active filter under varying network frequency. In: Proceedings of the 17th IFAC World Congress, pp. 3344–3349 (2008)
3. Fujioka, H.: A discrete-time approach to stability analysis of systems with aperiodic sample-and-hold devices. IEEE Transactions on Automatic Control 54(10), 2440–2445 (2009)
4. Hanson, R.D., Tsao, T.-C.: Periodic sampling interval repetitive control and its application to variable spindle speed noncircular turning process. Journal of Dynamic Systems, Measurement, and Control 122(3), 560–566 (2000)
5. Hillerström, G.: On Repetitive Control. PhD thesis, Lulea University of Technology (November 1994)
6. Robert, D., Sename, O., Zsimon, D.: Synthesis of a sampling period dependent controller using LPV approach. In: Proceedings of the 5th IFAC Symposium on Robust Control Design, Toulouse, France (2006)
7. Robert, D., Sename, O., Zsimon, D.: An H_∞ LPV design for sampling varying controllers: experimentation with a T inverted pendulum. IEEE Transactions on Control Systems Technology 18(3), 741–749 (2010)
8. Sala, A.: Computer control under time-varying sampling period: An LMI gridding approach. Automatica 41(12), 2077–2082 (2005)
9. Sánchez-Peña, R.S., Sznaier, M.: Robust Systems Theory and Applications. Adaptive and Learning Systems for Signal Processing, Communications and Control Series. Wiley-Interscience (August 1998)
10. Suh, Y.S.: Stability and stabilization of nonuniform sampling systems. Automatica 44(12), 3222–3226 (2008)
11. Tsao, T.-C., Qian, Y.-X., Nemani, M.: Repetitive control for asymptotic tracking of periodic signals with an unknown period. Journal of Dynamic Systems, Measurement, and Control 122(2), 364–369 (2000)
12. Weiss, G., Häfele, M.: Repetitive control of MIMO systems using H_∞ design. Automatica 35(7), 1185–1199 (1999)

Part II

HORC Approach

5
Odd-Harmonic High Order Repetitive Control

Summary. HORC is mainly used to improve the repetitive control performance robustness under disturbance/reference signals with varying or uncertain frequency. Unlike standard repetitive control, the HORC involves a weighted sum of several signal periods. With a proper selection of the associated weights, this high order function offers a characteristic frequency response in which the high gain peaks located at harmonic frequencies are extended to a wider region around the harmonics. Furthermore, the use of an odd-harmonic internal model will make the system more appropriate for applications where signals have only odd-harmonic components, as in power electronics systems. This Chapter presents an Odd-harmonic High Order Repetitive Controller suitable for applications involving odd-harmonic type signals with varying/uncertain frequency. The open loop stability of internal models used in HORC and the one presented here is analysed. Additionally, as a consequence of this analysis, an anti-windup scheme for repetitive control is proposed.

5.1 Introduction

5.1.1 State of the Art

The aim of HORC has been focused on two aspects: either improving the interharmonic behaviour by reducing the gain of the sensitivity function at these frequency intervals, or providing robustness against frequency variation/uncertainty by enlarging the interval around the harmonics for which the IM provides high gain.

The IM used in HORC can be defined from the generic IM (2.4) setting $\sigma = 1$ and

$$W(z) = \sum_{l=1}^{M} w_l z^{-lN},\qquad(5.1)$$

this yielding:

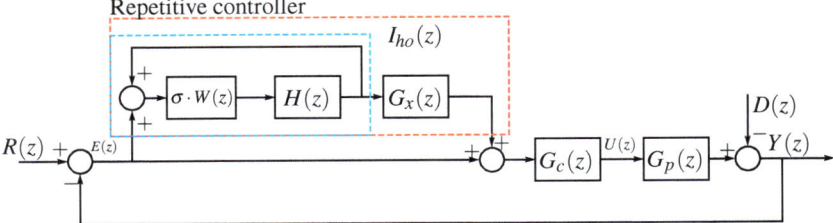

Fig. 5.1 Block-diagram of the repetitive controller plug-in approach

$$I_{ho}(z) = \frac{\sum_{l=1}^{M} w_l z^{-lN} H(z)}{1 - \sum_{l=1}^{M} w_l z^{-lN} H(z)}. \tag{5.2}$$

It can be seen that the delay function is now a weighted addition of M delays. The controller architecture follows the scheme depicted in Figure 5.1. Stability conditions for the closed-loop system of Figure 5.1 using (5.2) as the IM are derived in the same way as conditions given in Section 2.1. Thus, conditions 1 and 2 in Proposition 2.1 need to be fulfilled.

The weights w_l in (5.1) can be selected according to the desired performance. Aiming at this purpose, several criteria have been introduced which are usually formulated through an optimization procedure. In order to assure high gain at harmonic frequencies, the following constraint is commonly demanded:

$$\sum_{l=1}^{M} w_l = 1. \tag{5.3}$$

In case that $H(z) = 1$, condition (5.3) guarantees that the poles of the internal model (5.2) are located at harmonic frequencies.

In HORC, the robustness analysis against frequency variations is usually based on the modifying sensitivity transfer function of the system, $S_{Mod}(z)$ (see equation (2.9)).

In [11], HORC is used to improve the non-harmonic performance through the following optimization problem: $\min_{w_l} \|S_{Mod}(z)\|_2$; its analytical solution is given by $w_l = 1/M$, $\forall\, l$. In [1], the issue of finding the weights w_l is solved by means of an error spectrum minimization via $\min_{w_l} \|S_{Mod}(z)\|_\infty$. As a result, [11] and [1] do not amplify much interharmonic frequencies, but they do not improve robustness against frequency variations.

In [19], in order to minimize sensitivity against frequency variations, $W(z)$ is selected with the first $M-1$ derivatives equal to zero at harmonic frequencies, and an analytical solution is obtained. A generalization of the results in [1] is found in [20], where $\min_{w_l} \|G(z) S_{Mod}(z)\|_\infty$ is solved using the weight function $G(z)$ as a way to determine the frequency region for which $S_{Mod}(z)$ is to be minimized; results in [1, 19] can be obtained using this generic formulation.

In [15], the constraint (5.3) is eliminated. This diminishes the gain obtained at harmonic frequencies. Although this reduces the performance at nominal frequency,

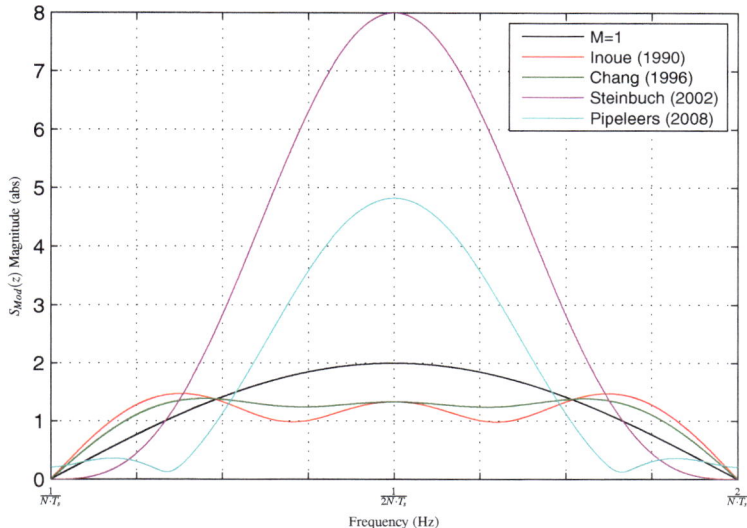

Fig. 5.2 Magnitude response of $S_{Mod}(z)$: comparison of [11], [1], [19] and [15] for $G_x(z) = 1/(G_o(z))^{-1}$, $H(z) = 1$ and $M = 3$

it makes possible the obtention of a lower interharmonic gain. In this way, the proposed optimization problem considers both components. It can be shown that the results in [1, 19, 20] can be obtained under specific settings of this formulation.

Figure 5.2 compares the results obtained for $G_x(z) = 1/(G_o(z))^{-1}$, $H(z) = 1$ and $M = 3$ when using the different tuning techniques previously introduced. The obtained weights are summarized in Table 5.1. Notice that [1, 11] enhance the interharmonic performance, but do not improve robustness against frequency variations. It is also shown that results from [19] yield perfect tracking, that is, the modifying sensitivity function gain at the harmonic frequencies is zero, and at the same time flatten the zone around them, which reflects the robustness improvement against small variation in the signal frequency. However, the frequency interval between the harmonics is notoriously amplified. Finally, although the result in [15] has no perfect tracking the gain around harmonics is kept small and the interharmonic amplification is lower than the one obtained in the previous case.

It is worth noticing that having too much gain at interharmonic frequencies can amplify undesirable signals with frequency components at those regions, such as noise or flicks, thus compromising the performance of the system.

5.1.2 Contribution

The main contribution of this part is based on proposing an odd-harmonic HORC. This controller is intended for systems that are exposed to periodic signals with only odd harmonic components, like those ones in power electronics applications, aiming at providing them with robustness against frequency variations. Two additional issues have been developed to complete the proposal: the stability analysis of the internal models used in HORC and, viewed as consequence of this, an anti-windup synthesis for repetitive control.

5.1.3 Outline

This Chapter is organized as follows. Section 5.2 studies the stability of the internal models used in HORC, which is carried out by means of the internal model pole analysis. Section 5.3 develops the Odd-harmonic High Order Repetitive Controller which constitutes the main matter of this Chapter. Section 5.4 proposes an anti-windup synthesis as a response to the analysis presented in Section 5.2 and, finally, conclusions and future work are presented in Section 5.5.

5.2 Internal Model Poles Analysis

In this section, the stability of the internal models used in HORC is analysed. This is an important issue since the practical implementation of the controller becomes more complex in case of unstable internal models. The stability of the models referenced in Section 5.1.1 is analysed and we are particularly interested in the internal models proposed in [19] and [15] since these approaches provide robustness in the face of frequency variation/uncertainty.

Proposition 5.1. *The weights obtained in [19] yields*

$$W(z) = 1 - (1 - z^{-N})^M$$

and, as a consequence, the IM resulting from (2.4) with $\sigma = 1$ and $H(z) = 1$ is

$$I_{ho}(z) = \frac{1 - (1 - z^{-N})^M}{(1 - z^{-N})^M}, \quad (5.4)$$

its poles being $z = \sqrt[N]{1}$ with multiplicity M.

Proof. By straightforward calculation. □

Table 5.1 Obtained weights using the proposals [11], [1], [19] and [15] $M = 3$

Delay blocks	$M = 3$
Inoue (1990)	$w_1 = w_2 = w_3 = 1/3$
Chang (1996)	$w_1 = 3/6, w_2 = 2/6, w_3 = 1/6$
Steinbuch (2002)	$w_1 = 3, w_2 = -3, w_3 = 1$
Pipeleers (2008)	$w_1 = -1.654588, w_2 = 1.521402, w_3 = -0.654055$

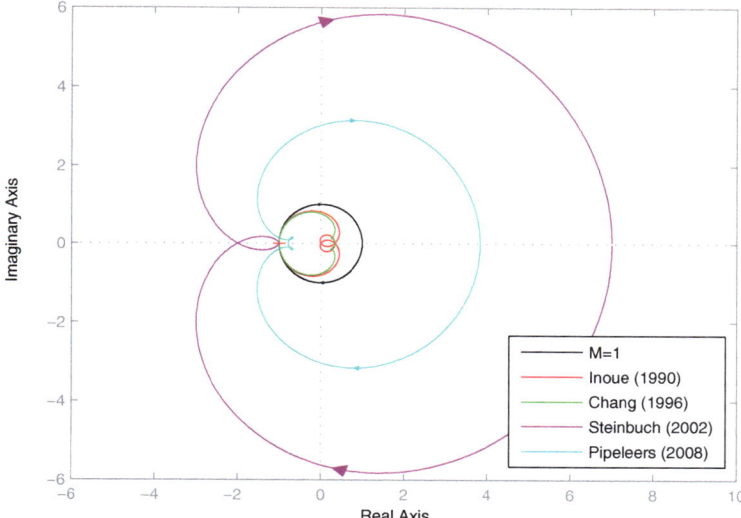

Fig. 5.3 Nyquist plot of $-W(z)$ with $N = 400$ for traditional RC ($M = 1$) and HORC ($M = 3$) tuned according to [11], [1], [19] and [15]

From Proposition 5.1 it is immediate that the poles coincide with those of the traditional repetitive controller ($M = 1$).

The increase of the multiplicity of the poles improves robustness against frequency variations [19] (i.e. with $M > 1$) but it makes the real implementation harder. For example, a system with actuator saturation in which the multiplicity of the poles located on the unit circle is greater than 1, will undergo a substantially harder wind-up effect.

Figure 5.3 shows the Nyquist plot of $-\sigma W(z)H(z)$ with $\sigma = 1$, $H(z) = 1$ and $N = 400$ for the options previously analyzed[1]. The Nyquist plot of the standard repetitive controller, i.e. with $M = 1$, is over the unit circle and, therefore, it is marginally stable[2]. It can be noticed that the results obtained in [1, 11] which do not improve robustness under frequency uncertainty, generate a Nyquist plot contained inside the unit circle, except at tangential points corresponding to the harmonic frequencies poles, whilst those methods which improve robustness [19], encircle the $(-1, 0)$ point many times.

Remark 5.1. Although, as shown in Proposition 5.1, the internal models obtained following the procedure in [19] for $H(z) = 1$ do not show poles outside the unit circle, with the introduction of a low pass filter $H(z)$ inside the IM the Nyquist plot

---

[1] Note that, as the internal model is composed of a positive feedback, the Nyquist criterion has to be applied to $-W(z)H(z)$.

[2] Marginally stable in the sense that it has poles on the unit circle with multiplicity one.

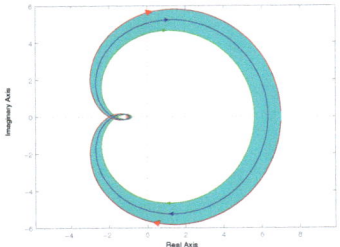
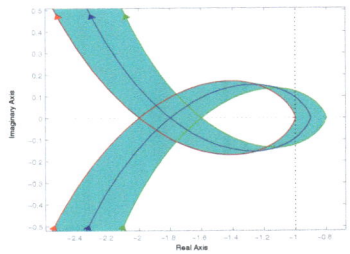

Fig. 5.4 Nyquist plot of $-W(z)H(z)$ for HORC ($M=3$) tuned according to [19] and using $H(z) = 0.05z^{-1} + 0.9 + 0.05z$

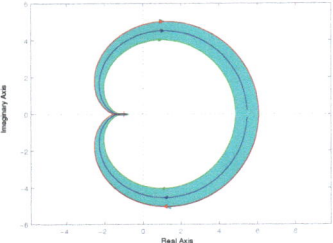
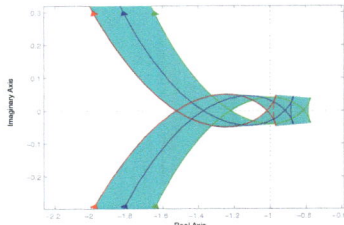

Fig. 5.5 Nyquist plot of $-W(z)H(z)$ for HORC ($M=3$) tuned according to [15] and using $H(z) = 0.05z^{-1} + 0.9 + 0.05z$

will vary slightly. The use of the traditional design of $H(z)$ causes this variation to change the number of the encirclements around $z = -1$ and, consequently, poles outside the unit circle appear in most cases.

As an example, Figure 5.4 shows the Nyquist plot of $-W(z)H(z)$ with $H(z) = 0.05z^{-1} + 0.9 + 0.05z$, $N = 400$ and $M = 3$, for the approach in [19]. As it can be seen, the addition of the filter $H(z)$ increases encirclements around $z = -1$ and the internal model has poles outside the unit circle.

The optimization-based tuning [15] generates internal models with poles outside the unit circle depending on the selected filter $H(z)$ and weight function. It is important to remark that, for the specific tuning shown in Figure 5.2, these internal models do not contain integrators or poles at the harmonic frequencies which generate an internal model without poles outside the unit circle. As an example, Figure 5.5 depicts the Nyquist plot of $-W(z)H(z)$ with $H(z) = 0.05z^{-1} + 0.9 + 0.05z$, $N = 400$ and $M = 3$, for the approach in [15] but using a different setup from the previous one. As it can be seen, the addition of the filter $H(z)$ and the new setup causes now encirclements around $z = -1$ and the resulting internal model has poles outside the unit circle.

From the theoretical point of view, the existence of poles outside the unit circle in the internal model does not compromise the closed-loop stability; however, unstable controllers yield an unpredictable behaviour in practical implementations where nonlinearities, actuator saturation or sensor failure can be present. As an example, sophisticated AW schemes must be included in order to avoid issues with saturated control actions [5]. Furthermore, regarding the interharmonic behaviour, an additional problem arises due to the water-bed effect which becomes harder for unstable open loop systems [18], such as those obtained with these internal models.

5.3 Odd-Harmonic HORC

5.3.1 Odd-Harmonic Repetitive Control

From the generic function (2.4) the standard IM [11] can be defined using $W(z) = z^{-N}$, with N being the discrete period of the signal, $H(z) = 1$ and $\sigma = 1$. This IM provides infinite gain at a certain fundamental frequency and all its harmonics until the $(N/2 - 1)$-th. Alternatively, an odd-harmonic IM, which only introduces infinite gain at odd-harmonics of the fundamental frequency [8], can be obtained with $W(z) = z^{-N/2}$, $H(z) = 1$ and $\sigma = -1$, yielding

$$I_{odd}(z) = \frac{-H(z)}{z^{\frac{N}{2}} + H(z)}. \tag{5.5}$$

When $H(z) = 1$, the poles of (5.5) are uniformly distributed over the unit circle[3]: $z = \exp(2(2k-1)\pi j/N)$, providing infinite gain at frequencies $\omega_k = 2(2k-1)\pi/N$, with $k = 1, 2, ..., N/2$.

The closed-loop system poles are the poles of the system without repetitive controller, i.e. the poles of $S_o(z)$ and the poles of $S_{Mod}(z)$, represented in equations (2.8) and (2.9), respectively. For the case in which $G_x(z) = k_r(G_o(z))^{-1}$, $\sigma = -1$, $W(z) = z^{-\frac{N}{2}}$ and $H(z) = 1$, the poles of $S_{Mod}(z)$ result in:

$$z = \sqrt[\frac{N}{2}]{|1-k_r|} e^{j(2\frac{(2k-1)\pi}{N} + \pi \cdot \frac{1-\text{sgn}(1-k_r)}{2})}, \; k = 1, \ldots, \frac{N}{2}. \tag{5.6}$$

These poles are uniformly distributed over a circle of radius $\sqrt[N/2]{|1-k_r|}$. To accomplish stability these poles should be within the unit circle, namely $k_r \in (0, 2)$. Although the introduction of $H(z) \neq 1$ and designs that involve nonminimum-phase plants affect the location of the closed-loop poles and also the convergence speed of the system as a function of k_r, this analysis gives a simple and intuitive approximation of the distribution of the poles [23].

---

[3] Note that, as there is no pole in $z = 1$, there is no infinite gain in DC-frequency, i.e. no integrator.

5.3.2 Odd-Harmonic HORC Internal Model

All previous works in HORC have been formulated for generic internal models providing full harmonic actuation. However, using the generic model (2.4), they can be transformed into odd-harmonic internal models replacing $\sigma = 1$ by $\sigma = -1$, N by $N/2$ and reformulating w_l in equation (5.1). Thus, the resulting high order odd-harmonic IM is:

$$I_{hodd}(z) = \frac{-W(z)H(z)}{1 + W(z)H(z)}, \quad (5.7)$$

with

$$W(z) = \sum_{l=1}^{M} (-1)^{l-1} w_l z^{-lN/2}. \quad (5.8)$$

Then, the computation of w_l based on the procedure presented in [19] can be carried out by forcing $W(e^{j\omega}) = -1$ at odd-harmonic frequencies, thus providing of perfect asymptotic tracking at those frequencies, and forcing the $M - 1$ first derivatives of $W(e^{j\omega})$ with respect to ω to be zero at odd-harmonic frequencies.

Similarly to [19], the procedure described here uses the maximally flat concept to calculate the weights of the function $W(z)$ in order to improve the robustness performance of the system.

It can be seen that the transfer function (5.7), with $H(z) = 1$, yields infinite gain when $W(z) = -1$. In the frequency domain that means

$$W(e^{j\omega}) = \sum_{l=1}^{M} (-1)^{l-1} w_l e^{-j\omega lN/2} = -1. \quad (5.9)$$

Since it is desirable to obtain infinite gain at odd-harmonic frequencies, it is required to set $\omega = 2\pi(2k-1)/N$ with $k = 1, 2, 3, \ldots$ in (5.9), which yields the following condition

$$\sum_{l=1}^{M} w_l = 1. \quad (5.10)$$

This constraint allows the achievement of perfect asymptotic tracking or disturbance rejection and guarantees that if the external signal is N-periodic with odd-harmonic content, the resulting weighted sum in (5.8) is the same as that obtained using just one delay element.

To show how the derivatives of $W(e^{j\omega})$ and $I_{hodd}(e^{j\omega})$ with respect to ω are related, the chain rule can be used: thus, for the first derivative one has

$$\frac{\partial I_{hodd}(e^{j\omega})}{\partial \omega} = \frac{\partial I_{hodd}(e^{j\omega})}{\partial W(e^{j\omega})} \frac{\partial W(e^{j\omega})}{\partial \omega} = \frac{-1}{(1+W(e^{j\omega}))^2} \frac{\partial W(e^{j\omega})}{\partial \omega},$$

while for the second derivative it is

$$\frac{\partial^2 I_{hodd}(e^{j\omega})}{\partial \omega^2} = \frac{\partial^2 I_{hodd}(e^{j\omega})}{\partial W^2(e^{j\omega})} \left(\frac{\partial W(e^{j\omega})}{\partial \omega}\right)^2 + \frac{\partial I_{hodd}(e^{j\omega})}{\partial W(e^{j\omega})} \frac{\partial^2 W(e^{j\omega})}{\partial \omega^2}$$

$$= \frac{2}{(1+W(e^{j\omega}))^3} \left(\frac{\partial W(e^{j\omega})}{\partial \omega}\right)^2 + \frac{-1}{(1+W(e^{j\omega}))^2} \frac{\partial^2 W(e^{j\omega})}{\partial \omega^2}.$$

The expressions to obtain higher order terms can be found using the Faà di Bruno's formula [3]. This result is useful to conclude that setting to zero the derivatives of $W(e^{j\omega})$ with respect to ω also sets to zero the derivatives of $I_{hodd}(e^{j\omega})$ with respect to ω.

Remark 5.2. It can be noticed that making $W(e^{j\omega})$ maximally flat at odd-harmonic frequencies also makes $I_{hodd}(e^{j\omega})$ maximally flat at those frequencies, thus increasing the frequency interval for which the internal model (5.7) provides the desired high gain.

Hence, to define the delay function $W(z)$, the weightings w_l can be calculated using (5.10) and making the first $M-1$ derivatives of $W(e^{j\omega})$ equal to 0 at odd-harmonic frequencies.

Thus, the first derivative is

$$\frac{\partial W(e^{j\omega})}{\partial \omega} = \sum_{l=1}^{M}(-1)^{l-1}w_l(-jl\frac{N}{2})e^{-j\omega lN/2}. \tag{5.11}$$

The condition states that

$$\left.\frac{\partial W(e^{j\omega})}{\partial \omega}\right|_{\omega=\frac{2(2k-1)\pi}{N}} = 0,$$

which gives

$$\sum_{l=1}^{M} w_l l = 0. \tag{5.12}$$

Also, the n−th derivative results:

$$\frac{\partial W(e^{j\omega})}{\partial \omega} = \sum_{l=1}^{M}(-1)^{l-1}w_l l^n(-j\frac{N}{2})^n e^{-j\omega lN/2}. \tag{5.13}$$

Thus, making the $M-1$ derivatives equal to 0 at $\omega = 2\pi(2k-1)/N$, the following compact condition is obtained:

$$\sum_{l=1}^{M} w_l l^p = 0. \tag{5.14}$$

with $p = 1, 2, \ldots, M-1$.

Hence, for $M = 3$, (5.10) yields $w_1 + w_2 + w_3 = 1$ and (5.14) yields $w_1 + 2w_2 + 3w_3 = 0$, $w_1 + 4w_2 + 9w_3 = 0$, which renders $w_1 = 3$, $w_2 = -3$, and $w_3 = 1$. In the same way, one gets $w_1 = 2$ and $w_2 = -1$ for $M = 2$.

The procedure described here attains the same conditions found in [19]. Also, the weights derived for HORC in [1, 15, 20] can be straightforwardly used for odd-harmonic HORC using definition (5.8). At the same time, the properties obtained from each method are preserved.

Equations (5.10) and (5.14) can be put together in the following compact form:

$$\sum_{l=1}^{M} l^p w_l = \begin{cases} 1 \text{ if } p=0 \\ 0 \text{ if } p=1,\ldots,M-1. \end{cases} \tag{5.15}$$

Proposition 5.2. *The weights obtained from (5.15) yield*

$$W(z) = -1 + \left(1 + z^{-\frac{N}{2}}\right)^M \qquad (5.16)$$

and, as a consequence, the IM resulting from (2.4) with $\sigma = -1$ *and* $H(z) = 1$ *is*

$$I_{hodd}(z) = \frac{1 - \left(1 + z^{-\frac{N}{2}}\right)^M}{\left(1 + z^{-\frac{N}{2}}\right)^M}, \qquad (5.17)$$

its poles being $z = \sqrt[N/2]{-1}$ *with multiplicity* M.

Proof. By straightforward calculation. □

In the light of Propositions 5.2 and 5.1 and using the same reasoning as before, it is immediate that the odd-harmonic internal models that can be obtained adapting the procedures [19] and [15] would yield the same (open loop) stability issues as those in the original version.

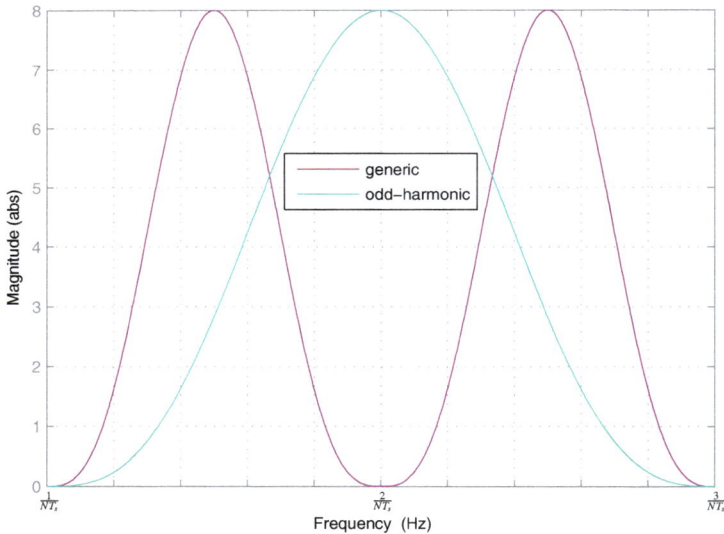

Fig. 5.6 Sensitivity functions of generic (Steinbuch (2002)) and odd-harmonic high-order repetitive controllers with $M = 3$ and $H(z) = 1$

Finally, it is worth remarking that although odd-harmonic internal models only introduce high gain at odd-harmonic frequencies, they do not improve the interharmonic behavior. In Figure 5.6 a comparison of the sensitivity function $S_{Mod}(z)$, for the generic [19] and the odd-harmonic internal models is presented using $H(z) = 1$ and $M = 3$. It can be seen that the maximum resulting gain at interharmonic frequencies is equal in both cases. However, robustness improvement against frequency variations/uncertainty at odd-harmonic frequencies is larger for the odd-harmonic IM presented here.

5.3.3 Selection of the Gain k_r

Besides the repetitive controller speed convergence, the parameter k_r affects the magnitude response of the system sensitivity function at the inter-harmonic frequency intervals. This is an important fact since HORC increases the gain at those intervals and some alleviation of that problem can be attained by a proper selection of k_r.

The modifying sensitivity function for the odd-harmonic HORC system is:

$$S_{Mod}(z) = \frac{1 + W(z)H(z)}{1 - W(z)H(z)(G_x(z)G_o(z) - 1)}. \quad (5.18)$$

The transfer function $S_{Mod}(e^{j\omega})$ is periodic in the frequency domain with period $4\pi/N$ under the assumption that $H(z) = 1$, $G_x(z) = k_r G_o^{-1}(z)$ and $W(z)$ as defined in (5.16). Thus, the magnitude response between two harmonics can be described from $S_{Mod}(e^{j\omega})$ using the normalized frequency $\bar{\omega} = \omega N/2$ with $\bar{\omega} \in [\pi, 3\pi]$:

$$\left| S_{Mod}(e^{2j\bar{\omega}/N}) \right| = \left| \frac{1 + W(e^{2j\bar{\omega}/N})}{1 - (k_r - 1)W(e^{2j\bar{\omega}/N})} \right|. \quad (5.19)$$

Figure 5.7, shows the magnitude of $S_{Mod}(z)$ for $M = 2$, $M = 3$ and for several values of k_r. It can be seen that $k_r < 1$ can be used to alleviate the over-amplification of frequencies between odd-harmonics while $k_r > 1$ causes an amplification of those frequencies.

Another important criterion used to select k_r is the closed-loop poles distance to the origin. The main idea around this criterion is to analyze the modulus of the closed-loop poles in order to obtain some insight about the convergence speed of the RC. As stated before, the closed loop poles are the poles of the sensitivity function $S(z)$ in equation (2.7) which, in turn, are the poles of $S_o(z)$ and $S_{Mod}(z)$. Thus, to analyze the speed convergence of the RC we focus our interest on the poles of $S_{Mod}(z)$. With $\sigma = -1$, $H(z) = 1$ and $G_x(z) = k_r/G_o(z)$, the closed-loop poles are the solutions of:

$$1 - W(z)(k_r - 1) = 0,$$

with $W(z) = z^{-N/2}$ and $W(z) = -1 + (1 + z^{-N/2})^M$ being the delay function for odd-harmonic RC and HORC, respectively.

In Figure 5.8, a comparison of the modulus of the closed-loop poles vs k_r is presented for odd-harmonic RC ($M = 1$), second order ($M = 2$) and third order ($M = 3$) HORC using $N = 100$. It can be concluded that the system speed convergence increases as k_r approaches to 1, and also that the increment in the RC order reduces the speed convergence.

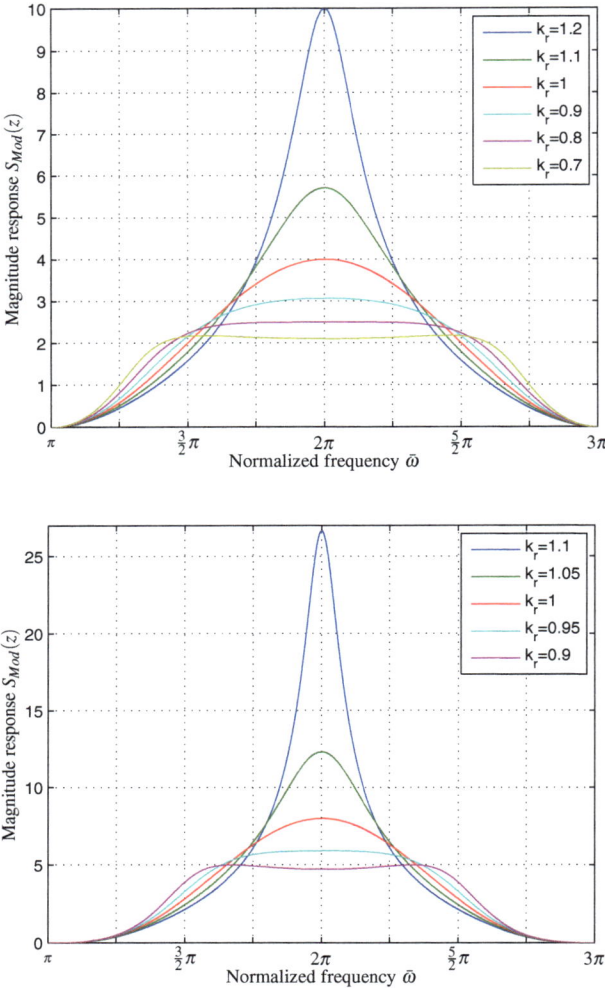

Fig. 5.7 $S_{Mod}(z)$ magnitude response for several values of k_r. Top: $M = 2$, bottom: $M = 3$.

5.3.4 Performance under Varying Frequency Conditions

Standard RC, including the odd-harmonic version, is designed assuming a constant period T_p. Therefore, if T_p varies the control algorithm performance may drastically decay. In order to illustrate the performance robustness of the odd-harmonic HORC proposed here, Figure 5.9 compares the magnitude response of the internal models used in odd-harmonic RC and odd-harmonic HORC.

Thus, Figure 5.9 highlights the gain of the internal models (5.5) and (5.7), designed for a nominal frequency of 50 Hz, for 49 Hz, 50 Hz and 51 Hz (and some

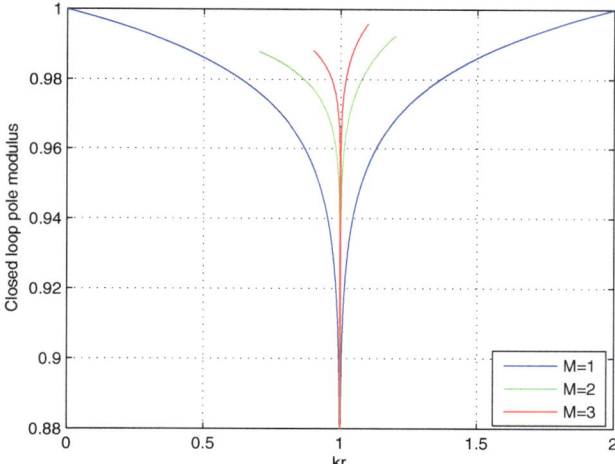

Fig. 5.8 Closed-loop poles modulus vs k_r for odd-harmonic RC ($M = 1$), second order ($M = 2$) and third order ($M = 3$) HORC, using $N = 100$

of their harmonics). On the one hand, the magnitude of the odd-harmonic function (5.5) is depicted in blue. Note that while for the 50 Hz signal the gain is important, it strongly decays for the other frequencies. On the other hand, the magnitude of the odd-harmonic HORC (5.7) for $M = 2$ and $M = 3$ is shown. It can be seen that the gain is higher for a wider frequency region around the harmonics, and this feature increases with the order of the internal model. This improves the robustness for variations in the period T_p. As a consequence, the gain decrease is much smaller for frequency variations around the nominal frequency in case of odd-harmonic HORC and the performance degradation is minor.

5.3.5 Second-Order Odd-Harmonic Internal Model

Two aspects make the second-order odd-harmonic IM an interesting case of HORC: firstly, the order of the IM and thus the need of memory in practical implementations is very similar to the standard RC with the additional benefit of providing robust performance and, secondly, the traditional design of the filter $H(z)$ renders an open-loop stable second-order IM.

In this section, sufficient conditions for the IM stability of second order odd-harmonic repetitive controllers are established. With this purpose, let us set $\sigma = -1$ and $M = 2$ in (5.16). Then, the IM obtained from (2.4) becomes:

$$I_{hodd}(z) = -\frac{\left(2z^{-\frac{N}{2}} + z^{-N}\right) H(z)}{1 + \left(2z^{-\frac{N}{2}} + z^{-N}\right) H(z)}. \quad (5.20)$$

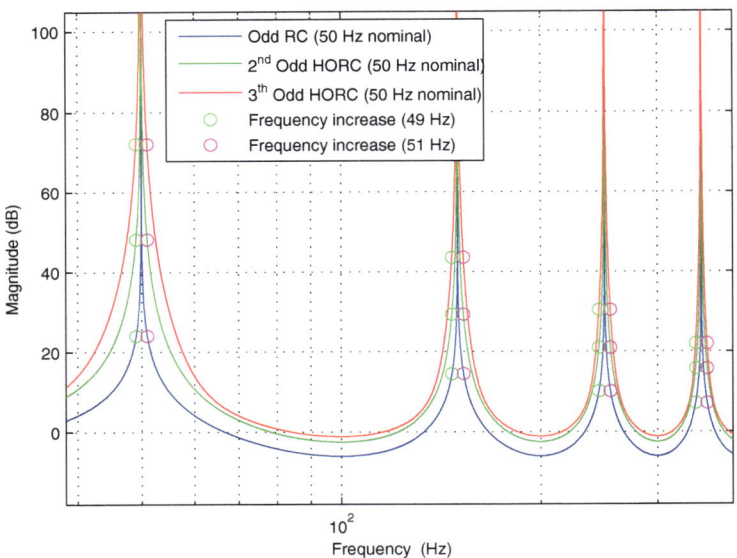

Fig. 5.9 Odd-harmonic RC and odd-harmonic HORC internal models gain diagram

Proposition 5.3. *The second-order, odd-harmonic IM (5.20), $H(z)$ being a null-phase filter with $\|H(z)\|_\infty < 1$, contains no poles outside the unit circle.*

Proof. Let us first show that (5.20) with $H(z) = 1$ has no poles outside the unit circle using a Nyquist-based argument. Notice that the Nyquist plot of $W(z)$ crosses the negative real axis at $\bar{\omega} = \frac{2(2 \cdot k+1)\pi}{N}$, $k = 0, \ldots, N/2 - 1$ and at $\bar{\omega} = \pi$ if N is multiple of 4. For all these frequencies the modulus of $W(e^{j\omega})$ is equal to 1. Hence, no encirclements are produced around $z = -1$, so the IM does not contain poles outside the unit circle.

Finally, the addition of a null-phase filter with $\|H(z)\|_\infty < 1$ does not modify the frequencies at which the negative real axis is crossed, since for all these frequencies:

$$|W(e^{j\bar{\omega}})H(e^{j\bar{\omega}})| \leq |W(e^{j\bar{\omega}})|\|H(z)\|_\infty < 1.$$

Thus, no encirclements around or crosses by $z = -1$ are produced and (5.20) is stable. □

Furthermore, assume that for the "plug-in" configuration of Figure 5.1:
(i) $G_x(z) = k_r (G_o(z))^{-1}$, this yielding the modifying sensitivity function obtained from (5.18) to become:

$$S_{Mod}^{hodd}(z) = \frac{1 + \left(2z^{-\frac{N}{2}} + z^{-N}\right) H(z)}{1 + \left(2z^{-\frac{N}{2}} + z^{-N}\right) H(z)(1 - k_r)}. \tag{5.21}$$

(ii) The poles of the closed-loop system without the repetitive controller, i.e. the poles of $S_o(z)$, are stable.

Proposition 5.4. *When $H(z) = 1$, the closed-loop system corresponding to the "plug-in" configuration of Figure 5.1 is stable for $k_r \in (0, \frac{4}{3})$.*

Proof. In accordance with the discussion in Section 5.3.1, the poles of the closed-loop system are given by those of $S_o(z)$ and $S_{Mod}^{hodd}(z)$. The poles of $S_o(z)$ are stable by hypothesis, while the poles of $S_{Mod}^{hodd}(z)$ obtained from (5.21) are

$$\bar{z} = \sqrt[\frac{N}{2}]{k_r - 1 \pm \sqrt{k_r^2 - k_r}}.$$

Then, the analysis of the modulus of the poles reveals that $|\bar{z}| \leq 1, \forall k_r \in (0, \frac{4}{3})$. □

Remark 5.3. For a given $k_r \neq 1$, the closed-loop poles obtained with the standard repetitive controller are twice faster than the ones obtained with the second-order odd-harmonic repetitive controller. When $k_r = 1$ all poles are in $z = 0$ for the first ($M = 1$) and second-order ($M = 2$) odd-harmonics internal models.

Proposition 5.5. *When $H(z)$ is a null-phase filter with $\|H(z)\|_\infty < 1$ and $|H(e^{j\omega})|_{\omega=0} = 1$, the closed-loop system corresponding to the "plug-in" configuration of Figure 5.1 is stable for $k_r \in (2/3, 4/3)$.*

Proof. As it is immediate that $\|W(z)\|_\infty = 3$, the result follows straightforward from the stability of $G_o(z)$ and the sufficient stability condition 2 in Proposition 2.1. □

In the light of Proposition 5.5 and taking different values for M, Figure 5.10 shows a comparison of the values of k_r for which the system is stable. It is worth to note that for higher values of M the k_r interval that assures system stability is highly reduced yielding a small margin for a successful practical implementation design.

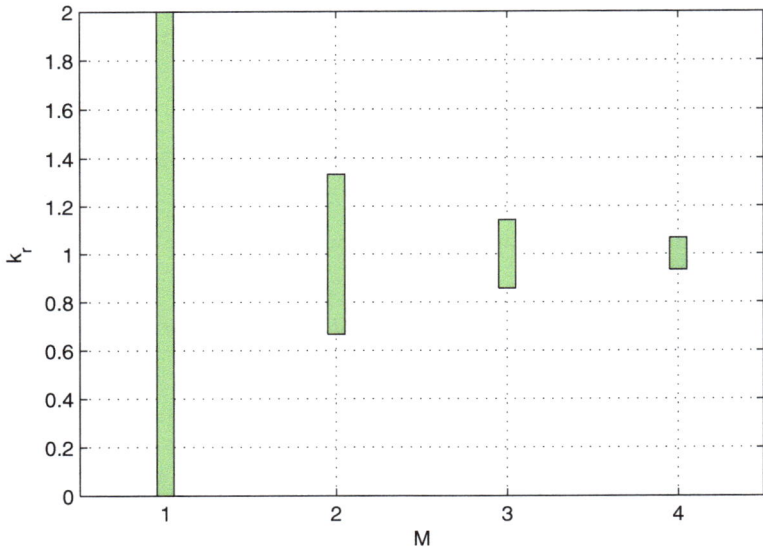

Fig. 5.10 Values of k_r vs. M for which the system is stable. $H(z)$ is a null-phase filter with $\|H(z)\|_\infty < 1$, $|H(e^{j\bar{\omega}})|_{\omega=0} = 1$ and $G_p(z)$ is a minimum-phase plant.

5.4 Anti-windup Synthesis for Repetitive Control

5.4.1 Introduction

Motivation

As discussed in previous chapters, the IM of the RC provides infinite or very high gain at a given frequency an its harmonics. It is well known that, in systems with actuator saturation, a controller with these features may produce a wind-up effect in which the states of the controller can grow unbounded [10]. Even if the gain is not infinite but high, the states can overgrow significantly, making it harder to recover the system to the linear ideal one. Some conditions related to the boundedness of the state of the RC with actuator saturation are stated in [17].

It is also known that marginally stable or unstable controllers are prone to originate an unbounded growing of the controller state. Thus, using the pole analysis presented in Section 5.2, it can be noted that the IM used in standard RC is marginally stable and those used in HORC [11, 19] have poles over the unit circle with multiplicity equal or greater than two, which can yield BIBO unstable IM. Additionally, the time response of the IM generally imposes a slow transient response for the closed loop which worsens the performance in the face of actuator saturation. Therefore, since the linear design of the RC does not take into account actuator saturation, the inclusion of an AW compensator is required.

Generic Anti-windup Designs

A recent review of standard AW techniques can be found in [5, 22]. In [5], modern AW proposals have been classified in two groups: Direct Linear Anti-Windup (DLAW) and Model Recovery Anti-Windup (MRAW). The DLAW approach seeks to find an AW compensator that assures specific performance and stability properties for the closed loop system. The MRAW approach selects the AW filter in such a way that it is a dynamical system containing the model of the plant. However, due to the characteristics of the RC, most of the standard AW designs should not be applied straightforward since some difficulties can arise during design or implementation, as discussed below. As a result, it would be necessary to adapt the generic AW strategies in order to be applied in RC.

In the DLAW scheme, some strategies are based on solving an LMI problem; however, issues usually arise since the size of this LMI depends mostly on the IM order, which is usually large. Thus, for the RC case, the implementation of this scheme depends on whether the LMI is computationally solvable or not. Although the DLAW scheme allows us to obtain an anti-windup compensator of order 0, the solution includes elements that yield a large number of on-line calculations, thus increasing the computational burden.

The MRAW scheme uses the model of the plant in its structure. Although the plant order could be large, it is usually significantly smaller than the IM order. Furthermore, the procedure to find the feedback gains does not depend on the controller dynamics. Therefore, the related LMI is always solvable. The computational load of the MRAW scheme implementation is the lowest one in comparison with the other strategies.

Anti-windup Designs for RC : State of the Art

The proposals in [4, 16, 17] are three examples of AW design for RC.

In [16], an AW law is obtained for Iterative Learning Control (ILC) and also a extension to RC is briefly described. However, the AW strategy is derived for a specific plant and the described repetitive controller does not correspond to the standard architecture since the filters for stability and robustness are not included. In [17], the AW scheme cancels out the dynamics of the IM during saturation and adds a structure to shape the transients when the system saturates and gets back from saturation. However, the IM cancellation implies that, in addition to the repetitive controller order, it is necessary to implement an AW filter which has at least the order of the IM. Therefore, this scheme will be cost restrictive since it depends on a suitable implementation platform.

The work in [4] can be categorized as a DLAW design. The strategy, derived in continuous-time domain, is an extension to RC of the general AW design in [2], where the case of delayed systems is described. Also in this approach, unlike the RC design that will be described here, the filters for robustness and stability are designed together with the DLAW synthesis.

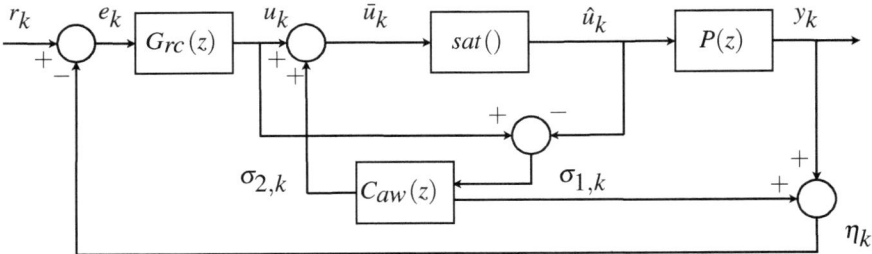

Fig. 5.11 The MRAW scheme in RC

Contribution

In view of the analysis above, the MRAW appears as a good strategy when taking into account the computational solvability and the computational load in the design and implementation of RC. Thus, in this work a MRAW scheme is presented in which the recovery of the system is achieved using the approximation to a deadbeat transition [14]. The advantage of selecting a deadbeat over other designs is shown, as well as the design method.

5.4.2 The General MRAW Scheme

Figure 5.11 shows the MRAW structure, where $G_p(z)$ is the plant, $G_{rc}(z)$ is the controller, $sat(\cdot)$ is the saturation function and $C_{aw}(z)$ is the AW compensator. In the MRAW strategy, the mismatch between the saturated control action and the non-saturated one is fed back to the controller by means of the AW compensator which is designed to be the model of the plant, $\sigma_{1,k}$ being the output that is used with this purpose. Additionally, another feedback signal, $\sigma_{2,k}$, is added aiming at improving the behavior of the system when it gets out from saturation. Thus, the design of this feedback involves different approaches. The IMC AW strategy [25], turns out to be the particular case where $\sigma_{2,k} = 0$. This causes that, when getting out of saturation, the system recovery relies on the plant poles, which can yield an unsuitable performance. A strategy Predictive Control-based which seeks an l_2 performance criterion can be found in [7], an optimization procedure using the LQ approach is proposed in [24] and a fully nonlinear strategy is described in [6].

In this Section, the signal $\sigma_{2,k}$ is designed to be a linear feedback of the AW compensator state. This is aimed at finding a simple linear solution to the AW problem in case of RC, also avoiding the algebraic loop that can be created using the feedback of the control action mismatch, as in [24]. Furthermore, we analyze the benefits of designing a deadbeat behavior in the AW filter in case of RC. Also, as previously mentioned, the advantage of using the MRAW scheme for the RC case is that the design does not depend on the IM order. Additionally, as will be described further on, the error and control signals are the ideal ones (as if the system had no saturation in the actuator), which isolates the controller from saturation effects.

Selected MRAW Scheme

Consider the MRAW scheme depicted in Figure 5.11. Let the discrete-time, asymptotically stable linear plant $G_p(z)$ be

$$\begin{aligned} x_{k+1} &= Ax_k + Bsat(\bar{u}_k) \\ y_k &= Cx_k, \end{aligned} \quad (5.22)$$

where

$$sat(\bar{u}_k) = \begin{cases} u_{min} & \bar{u}_k < u_{min}, \\ \bar{u}_k & u_{min} \leq \bar{u}_k \leq u_{max}, \\ u_{max} & \bar{u}_k > u_{max} \end{cases} \quad (5.23)$$

with $u_{min} < 0$ and $u_{max} > 0$.

The state-space representation of the repetitive controller $G_{rc}(z)$ is:

$$\begin{aligned} \bar{x}_{k+1} &= A_{rc}\bar{x}_k + B_{rc}e_k \\ u_k &= C_{rc}\bar{x}_k + D_{rc}e_k. \end{aligned} \quad (5.24)$$

The AW filter $C_{aw}(z)$ is defined from the plant model (5.22) as:

$$\begin{aligned} \chi_{k+1} &= A\chi_k + B(u_k - sat(u_k + \sigma_{2,k})) \\ \sigma_{1,k} &= C\chi_k \end{aligned} \quad (5.25)$$

and

$$\sigma_{2,k} = K\chi_k, \quad (5.26)$$

where K is the design parameter of the AW filter.

Notice that, while the input in system (5.22) is the saturated control action, the input in system (5.25) is the difference between the saturated and non-saturated control action. This fact, together with

$$\eta_k = y_k + \sigma_{1,k}, \quad (5.27)$$

helps to determine the system invariance. Thus, defining $\xi_k = x_k + \chi_k$, noticing that $\bar{u}_k = u_k + \sigma_{2,k}$ and adding equations (5.22) with (5.25) one has that:

$$\begin{aligned} \xi_{k+1} &= A\xi_k + Bu_k \\ \eta_k &= C\xi_k. \end{aligned} \quad (5.28)$$

In this way, from the input u_k to the output η_k, the system in Figure 5.12 can be seen as a LTI one with the dynamics of the plant.

This means that η_k is the ideal plant output in the sense that it would be the plant output in a system without actuator saturation. Furthermore, in the closed loop of Figure 5.11, the control action action u_k is the ideal control action, i.e. u_k is the same control signal as the one in a system without actuator saturation. This fact isolates the controller from the saturation effects, allowing us to reduce the analysis to the behaviour of the invariant part shown in Figure 5.12, including its internal stability.

Remark 5.4. In this scheme the deviation from the ideal performance can be measured through $\sigma_{1,k}$, since $\sigma_{1,k}$ is the difference between the ideal behaviour and the plant output $\sigma_{1,k} = \eta_k - y_k$.

Proposal 1. *Given a RC design, the smallest possible $\sigma_{1,k}$ corresponds to the best possible performance in case of saturation (the smallest deviation from the ideal behaviour). Therefore, the problem formulation is to find K such that $\sigma_{1,k}$ is small enough to obtain a good tracking performance.*

It is important that the AW design aims at achieving good tracking performance since RC is a technique which is intended to obtain null steady-state tracking error. Also due to this RC feature, we are interested in the saturation effect produced in steady state even though it can also occur in transient state.

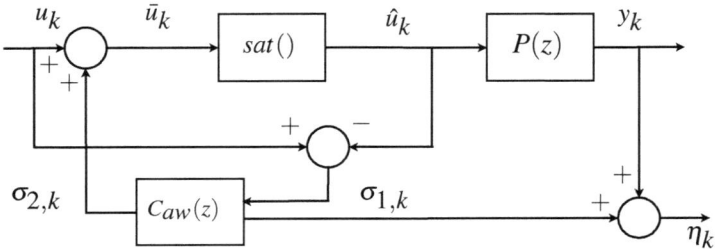

Fig. 5.12 The invariant part of the MRAW scheme

5.4.3 MRAW Proposal: The Deadbeat Anti-windup Controller for RC

This section analyses the selection of the design parameter K in the context of RC. The analysis is carried out in steady state, and therefore, in the RC frame we can consider that all the signals appearing in the system are periodic (assuming that non-periodic disturbance signals do not cause saturation we can neglect their effect on the system in this analysis). Furthermore, the actuator saturation depends on the signal to be tracked or rejected, i.e. the command signal or the disturbances.

In steady state, we can assume as standard scenario a control action that saturates only in specific parts of each period of the signal. Thus, at that instant the plant output deviates from the ideal behavior and in the other part of the period the system can recover.

The system saturates when $u_k + \sigma_{2,k} > u_{max}$, moment at which the input of the AW filter, the difference $u_{max} - u_k \neq 0$ and its dynamics corresponds with the plant dynamics, thus establishing the invariance from u_k to η_k. Notice that $\sigma_{2,k}$ only has influence in the saturation function condition $u_k + \sigma_{2,k} > u_{max}$, helping to determine when the system can go out from saturation. Thus, the design of K affects the amount of time the system remains saturated. Also, notice that K does not affect neither the magnitude of $\sigma_{1,k}$ nor the AW state dynamics χ_k at this stage.

5.4 Anti-windup Synthesis for Repetitive Control

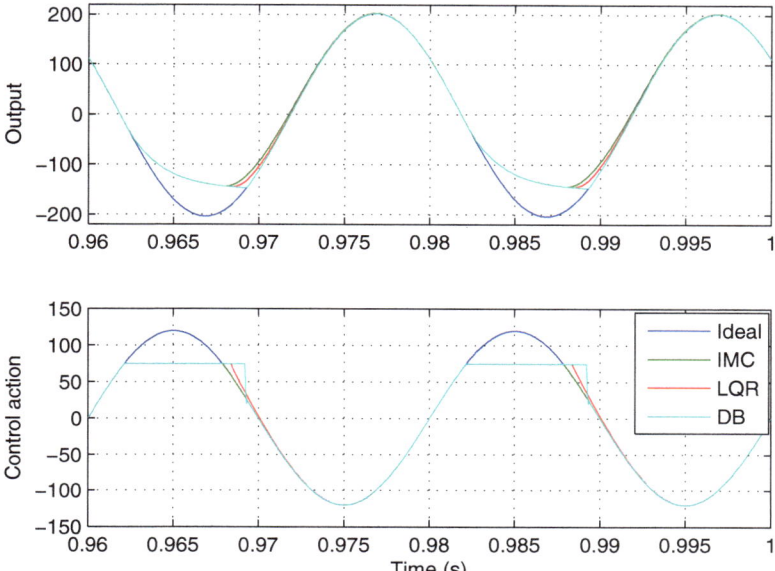

Fig. 5.13 AW compensator comparison: Ideal response for the system without saturation, IMC for $K = 0$, DB for a deadbeat behaviour and LQR for an intermediate behaviour

Getting out of saturation, this is when $u_k + \sigma_{2,k} < u_{max}$, the vanishing of the AW filter state χ_k, and so $\sigma_{1,k}$, will lie in the dynamics imposed by the state feedback loop created by $\sigma_{2,k}$. In this way, we can obtain the IMC strategy with $K = 0$, while also other dynamics can also be obtained using different pole placement designs. In general we are interested in values of K that provide $C_{aw}(z)$ with dynamics faster than the plant one.

Proposal 2. *The feedback gain K can be designed to be a deadbeat solution, thus locating the poles of $C_{aw}(z)$ in the origin and providing the fastest vanishing of the states and outputs of $C_{aw}(z)$.*

For more details about the deadbeat design and behavior see [14].

As an example[4], let us assume that the signal to be tracked is a sine waveform signal, and that the saturation occurs due to the amplitude of the sinusoidal control signal for $u \geq 75$. Figure 5.13 shows the ideal linear behavior and the response when there is saturation in the actuator and three different values of K: $K = 0$ i.e. IMC AW strategy, K for a deadbeat (DB) AW filter and K obtained from a LQR design with a faster dynamics than the plant one.

---

[4] The simulation has been performed using the scheme of Figure 5.12 with $u_k = 120 \cdot sin(100\pi t_k)$, sampling time $T_s = 50$ μs and the discrete time model of the active power filter in Chapter 7

In the first case, with $K = 0$ i.e. the IMC strategy where $\sigma_{2,k} = 0$, the saturated control action is simply the result of limiting the ideal control action to its maximum, namely the system only saturates when $u_k > u_{max}$. Thus, once the system returns to the linear zone, $u_{max} - u_k = 0$, the state of the AW filter, and so $\sigma_{1,k}$, will vanish with the plant model dynamics, whereupon the output of the plant will be the ideal one. Is it also worth to notice that this is an asymptotic behavior.

In the deadbeat case, once $u_k + \sigma_{2,k} < u_{max}$, the state of $C_{aw}(z)$ and its outputs, $\sigma_{1,k}$ and $\sigma_{2,k}$, will vanish in n sampling steps, where n is the relative degree of the plant model. At that moment $y_k = \eta_k$ (and $\psi = 0$). This means that, based on the system invariance, the system will remain saturated until the output plant y_k, with input u_{max}, can equal the ideal output η_k. Therefore, we are applying the maximum possible effort by means of u_{max}, until the moment the ideal input can be reached again.

For other values of K, the AW filter will exhibit an intermediate behavior with asymptotic response when getting out of saturation.

Comparing the response for different values of K we have that during saturation all the systems have similar behaviour since K does not modify the value of $\sigma_{1,k}$; however, the amount of time the system saturates is different, the longest being the deadbeat design. Also, it can be seen that the system which remains saturated longer has a smaller $\sigma_{1,k}$.

5.4.4 MRAW Proposal: Design and Stability

The proposal is based on the idea of having a deadbeat recovery once the system gets back from saturation. The goal is to obtain a $C_{aw}(z)$ AW filter such that during saturation it takes the form of the plant model, and additionally, when the control action gets back from saturation, the outputs of $C_{aw}(z)$ vanish in a finite number of samples.

To obtain a deadbeat behavior during recovery it is needed that the feedback loop created by $\sigma_{k,2}$ relocates all the poles of $C_{aw}(z)$ to $z = 0$, which can be done using the pole placement procedure, thus obtaining the gains vector K. However, the internal stability of the system must be verified.

5.4.5 Stability

Remark 5.5. The closed loop stability of the system in Figure 5.11 is established by the design of the RC and, additionally, by the internal stability of the system in Figure 5.12.

Moreover, from the fact that: 1) we are assuming an asymptotically stable plant and 2) from input u_k to output η_k the system in Figure 5.12 can be seen as a LTI one with the dynamics of the plant, we have that the internal stability of this system can be established just analyzing the stability of the interconnection between the saturation block and $C_{aw}(z)$ (see Figure 5.14).

5.4 Anti-windup Synthesis for Repetitive Control

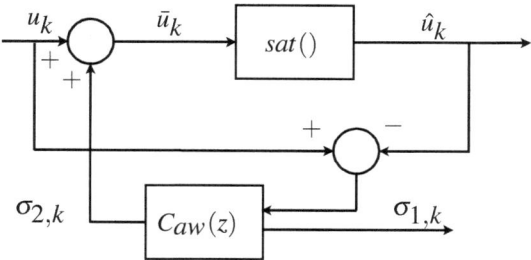

Fig. 5.14 The connection between the AW compensator $C_{aw}(z)$ and the saturation block

Proposition 5.6. *Given an open loop stable AW compensator $C_{aw}(z)$, as defined in (5.25), the closed loop system in Figure 5.14 is $\mathscr{L}_2$ stable from the input u_k to the state χ_k of $C_{aw}(z)$ if there exist $P = P^T > 0$ and $W = W^T > 0$ such that:*

$$\begin{bmatrix} A^T PA - P & -A^T PB + K^T W \\ -B^T PA + WK & -2W + B^T PB \end{bmatrix} < 0. \tag{5.29}$$

Proof. In a first step, it can be easily shown that the function $sat()$ satisfies the following inequality:

$$\Psi = 2(\bar{u}_k - sat(\bar{u}_k))W(sat(\bar{u}_k)) \geq 0, \tag{5.30}$$

with W obtained from (5.29). Noticing that $\bar{u}_k = u_k + K\chi_k$, equation (5.30) can be rewritten as:

$$\Psi = \bar{\Psi} + \tilde{\Psi},$$

where

$$\bar{\Psi} = sat(\bar{u}_k)WK\chi_k + \chi_k^T K^T W sat(\bar{u}_k) - 2sat(\bar{u}_k)W sat(\bar{u}_k)$$

and

$$\tilde{\Psi} = 2u_k W sat(\bar{u}_k).$$

In a second step, based on equation (5.25) and system interconnection in Figure 5.14 we propose the candidate Lyapunov function $V(\chi_k) = \chi_k^T P \chi_k$, where P is obtained from (5.29). Thus we find that:

$$\Delta V = V(\chi_{k+1}) - V(\chi_k) = (A\chi_k + Bu_k - Bsat(\bar{u}_k))^T P (A\chi_k + Bu_k - Bsat(\bar{u}_k)) - \chi_k^T P \chi_k,$$

which can be expressed as:

$$\Delta V = \bar{\Delta V} + \tilde{\Delta V},$$

where

$$\bar{\Delta V} = \chi_k^T (A^T PA - P)\chi_k - sat(\bar{u}_k)B^T PA\chi_k - \chi_k^T A^T PB sat(\bar{u}_k)$$
$$+ sat(\bar{u}_k)B^T PB sat(\bar{u}_k)$$

and
$$\tilde{\Delta V} = u_k B^T P A \chi_k + \chi_k^T A^T P B u_k \\ - u_k^T B^T P B sat(\bar{u}_k) - sat(\bar{u}_k) B^T P B u_k + u_k^T B^T P B u_k.$$

Hence, based on (5.30), ΔV satisfies the following inequality:
$$\Delta V \leq \bar{\Delta V} + \tilde{\Delta V} + \bar{\Psi} + \tilde{\Psi}.$$

It can be easily seen that (5.29) corresponds to:
$$\bar{\Delta V} + \bar{\Psi} < 0.$$

Since this is a strict inequality then there exists a small enough scalar $\varepsilon > 0$ such that
$$\bar{\Delta V} + \bar{\Psi} \leq -\varepsilon \left(\chi_k^T \chi_k + sat(\bar{u}_k)^2 \right).$$

Therefore, it results that
$$\Delta V + \varepsilon \left(\chi_k^T \chi_k + sat(\bar{u}_k)^2 \right) - \tilde{\Delta V} - \tilde{\Psi} \leq 0$$

Then, using completion of squares, we obtain:
$$\Delta V + \frac{\varepsilon}{2} \chi_k^T \chi_k + \frac{\varepsilon}{2} \left(\chi_k - \frac{2}{\varepsilon} A^T P B u_k \right)^T \left(\chi_k - \frac{2}{\varepsilon} A^T P B u_k \right) - u_k^T Q_1 u_k \\ + \varepsilon \left(sat(\bar{u}_k) + \frac{1}{\varepsilon} \left(B^T P B - W \right) u_k \right)^2 - u_k^T Q_2 u_k - u_k^T B^T P B u_k \leq 0,$$

with $Q_1 = \frac{2}{\varepsilon} B^T P A A^T P B$ and $Q_2 = \frac{1}{\varepsilon} \left(B^T P B - W \right)^2$. With this, the following holds
$$\Delta V + \frac{\varepsilon}{2} \chi_k^T \chi_k - u_k^T Q_T u_k \leq 0,$$

with $Q_T = \frac{2}{\varepsilon} B^T P A A^T P B + \frac{1}{\varepsilon} \left(B^T P B - W \right)^2 + B^T P B$. Therefore, there exists a sufficiently large $\gamma > 0$ such that:
$$\Delta V + \frac{\varepsilon}{2} \chi_k^T \chi_k \leq \gamma u_k^T u_k,$$

Adding up both sides of previous equation from 0 to ∞ we have
$$(V(\chi_\infty) - V(\chi_0)) + \frac{\varepsilon}{2} \sum_{k=0}^{\infty} \chi_k^T \chi_k \leq \gamma \sum_{k=0}^{\infty} u_k^T u_k;$$

then, assuming that the candidate Lyapunov function is such that, $V(\chi_\infty) = 0$ we have
$$\frac{\varepsilon}{2} \sum_{k=0}^{\infty} \chi_k^T \chi_k \leq \gamma \sum_{k=0}^{\infty} u_k^T u_k + \chi_0^T P \chi_0,$$

and $\mathscr{L}_2$ stability from u_k to χ_k follows. □

Inequality (5.30) is referred in the literature as the sector condition (see [12]). Thus, the memoryless function $sat()$ is said to belong to the sector $[0,1]$ since $sat(t,u)[u - sat(t,u)] \geq 0$.

Finally, it is important to note that (5.29) is a nonlinear inequality. Therefore, to make it linearly solvable we can apply the Schur–complement and a congruence transformation [9]. Thus, (5.29) can be expanded as:

$$\begin{bmatrix} -P & K^T W \\ WK & -2W \end{bmatrix} - \begin{bmatrix} A^T \\ -B^T \end{bmatrix} [-P] [A \ -B] < 0,$$

then, applying the Schur–complement we obtain

$$\begin{bmatrix} -P^{-1} & A & -B \\ A^T & -P & K^T W \\ -B^T & WK & -2W \end{bmatrix} < 0.$$

We can perform the following congruence transformation

$$\begin{bmatrix} I & 0 & 0 \\ 0 & P^{-1} & 0 \\ 0 & 0 & W^{-1} \end{bmatrix} \begin{bmatrix} -P^{-1} & A & -B \\ A^T & -P & K^T W \\ -B^T & WK & -2W \end{bmatrix} \begin{bmatrix} I & 0 & 0 \\ 0 & P^{-1} & 0 \\ 0 & 0 & W^{-1} \end{bmatrix} < 0,$$

resulting in

$$\begin{bmatrix} -P^{-1} & AP^{-1} & -BW^{-1} \\ P^{-1}A^T & -P^{-1} & -P^{-1}K^T \\ -W^{-1}B^T & KP^{-1} & -2W^{-1} \end{bmatrix} < 0,$$

then, using $Q = P^{-1}$, $U = W^{-1}$, $X_1 = KQ$ the following LMI is obtained

$$\begin{bmatrix} -Q & AQ & -BU \\ QA^T & -Q & X_1^T \\ -UB^T & X_1 & -2U \end{bmatrix} < 0. \qquad (5.31)$$

Hence, if the LMI (5.31) is feasible with $Q = Q^T > 0$ and $U = U^T > 0$ we can establish the system stability for the previously designed deadbeat gain K.

5.4.6 Optimal Design

In the above design a deadbeat feedback gain K is synthesized using a pole placement strategy. Nevertheless, it was shown that one needs to check the system internal stability after that procedure. In this section, following a different idea, we propose a design which looks for obtaining the deadbeat behavior previously detailed and, at the same time, assures global asymptotic stability. In this way we propose to put together condition (5.31) with an optimal LQ design in LMI form.

Remark 5.6. In this case the LQ design is used to find the deadbeat gain K which is shown to be optimal when the weight matrix $Q_p = T^T T$, T being the linear transformation of the system (5.25) into the controllable canonical form (see [13]).

Other results related with the deadbeat design as an optimal LQ solution can be found in [21]. Thus, the problem formulation is to find K such that the stability of the interconnection between $C_{aw}(z)$, equations (5.25) and (5.26), and the saturation block is preserved and additionally, to solve the constrained LQ problem:

$$\min_{K} \sum_{k=0}^{\infty} \chi_k^T Q_p \chi_k$$

subject to:

$$\chi_{k+1} = A\chi_k - B\sigma_{2,k}$$
$$\sigma_{2,k} = K\chi_k,$$

The complete problem can be formulated as an LMI minimization problem:

$$\min \gamma$$
s.t.
$$\begin{bmatrix} -Q & AQ & -BU \\ QA^T & -Q & X_1^T \\ -UB^T & X_1 & -2U \end{bmatrix} < 0,$$

$$\begin{bmatrix} \gamma I & I \\ I & -Q \end{bmatrix} > 0,$$

$$\begin{bmatrix} -Q & QA^T - X_1^T B^T & QQ_p \\ AQ - BX_1 & -Q & 0 \\ Q_p Q & 0 & -Q_p \end{bmatrix} < 0,$$

where $Q = Q^T > 0$, $U = U^T > 0$, $\gamma > 0$ and $X_1 = KQ$. It is worth to note that the first LMI corresponds to the stability condition (5.31) and the latter two LMIs are the optimal LQ design in LMI form (see appendix C for derivation of this LMI).

Also, it is worth pointing out that there exists some conservativeness in the sector condition Ψ which is applied to non-linearities belonging to the sector $[0,1]$. In general, this fact may yield a gain K that is an approximation to the deadbeat solution.

5.5 Conclusions

As the main contribution, this Chapter proposed an odd-harmonic HORC. It was shown that this controller provides robust performance in case of signals with uncertain or varying frequency. As a consequence, the proposed controller allows the system to operate in a wider frequency range without reducing its performance and without requiring a frequency observer or adaptive mechanisms. A comparison with an odd-harmonic repetitive controller revealed a better efficiency of the proposed controller working under varying frequency conditions. The stability of the internal models used in HORC was analyzed and a stable second order internal model for repetitive controllers was proposed and studied. Additionally, the implementation complexity turns out to be the same when comparing the order and computational cost of the standard (full-harmonic) repetitive controller and the second order odd-harmonic repetitive controller proposed here. Finally in the last Section, the Model Recovery AW scheme was studied and adapted to the RC case. An optimal LQ design was proposed aimed at finding a deadbeat recover behavior and assuring the global asymptotic stability of the closed loop system. The future research should study the inclusion of less-restrictive sector conditions for the nonlinear saturation function in order to better approximate the deadbeat design.

References

1. Chang, W.S., Suh, I.H., Kim, T.W.: Analysis and design of two types of digital repetitive control systems. Automatica 31(5), 741–746 (1995)
2. da Silva Jr., J.G., Tarbouriech, S.: Antiwindup design with guaranteed regions of stability: an LMI-based approach. IEEE Transactions on Automatic Control 50(1), 106–111 (2005)
3. di Bruno, F.F.: Note sur une nouvelle formule du calcul diffrentielle. The Quarterly Journal of Pure and Applied Mathematics 1, 359–360 (1857)
4. Flores, J., Gomes da Silva, J., Pereira, L., Sbarbaro, D.: Robust repetitive control with saturating actuators: a LMI approach. In: Proceedings of the American Control Conference (ACC), pp. 4259–4264 (July 2010)
5. Galeani, S., Tarbouriech, S., Turner, M., Zaccarian, L.: A tutorial on modern anti-windup design. European Journal of Control 15, 418–440 (2009)
6. Galeani, S., Teel, A.R., Zaccarian, L.: Constructive nonlinear anti-windup design for exponentially unstable linear plants. Systems & Control Letters 56(5), 357–365 (2007)
7. Grimm, G., Teel, A.R., Zaccarian, L.: The L2 anti-windup problem for discrete-time linear systems: Definition and solutions. Systems & Control Letters 57(4), 356–364 (2008)
8. Griñó, R., Costa-Castelló, R.: Digital repetitive plug-in controller for odd-harmonic periodic references and disturbances. Automatica 41(1), 153–157 (2005)
9. Herrmann, G., Turner, M., Postlethwaite, I.: Linear matrix inequalities in control. In: Turner, M., Bates, D. (eds.) Mathematical Methods for Robust and Nonlinear Control. LNCIS, vol. 367, pp. 123–142. Springer, Heidelberg (2007)
10. Hippe, P.: Windup in Control: Its Effects and Their Prevention. Advances in Industrial Control. Springer (2010)
11. Inoue, T.: Practical repetitive control system design. In: Proceedings of the 29th IEEE Conference on Decision and Control, pp. 1673–1678 (1990)
12. Khalil, H.K.: Nonlinear Systems, 3rd edn. Prentice Hall, Upper Saddle River (2002)
13. Kučera, V.: Deadbeat response is L2 optimal. In: Proceedings of the 3rd International Symposium on Communications, Control and Signal Processing, ISCCSP 2008, pp. 154–157 (March 2008)
14. Kučera, V.: Analysis and Design of Discrete Linear Control Systems. Prentice Hall (1991)
15. Pipeleers, G., Demeulenaere, B., Sewers, S.: Robust high order repetitive control: Optimal performance trade offs. Automatica 44, 2628–2634 (2008)

16. Ryu, Y.S., Longman, R.: Use of anti-reset windup in integral control based learning and repetitive control. In: Proceedings of the IEEE International Conference on Systems, Man, and Cybernetics. Humans, Information and Technology, vol. 3, pp. 2617–2622 (October 1994)
17. Sbarbaro, D., Tomizuka, M., de la Barra, B.L.: Repetitive control system under actuator saturation and windup prevention. Journal of Dynamic Systems, Measurement, and Control 131(4), 044505 (2009)
18. Seron, M., Goodwin, G.C., Braslavsky, J.: Fundamental limitations in filtering and control. Springer, London (1997)
19. Steinbuch, M.: Repetitive control for systems with uncertain period-time. Automatica 38(12), 2103–2109 (2002)
20. Steinbuch, M., Weiland, S., Singh, T.: Design of noise and period-time robust high order repetitive control, with application to optical storage. Automatica 43, 2086–2095 (2007)
21. Sugimoto, K., Inoue, A., Masuda, S.: A direct computation of state deadbeat feedback gains. IEEE Transactions on Automatic Control 38(8), 1283–1284 (1993)
22. Tarbouriech, S., Turner, M.: Anti-windup design: an overview of some recent advances and open problems. Control Theory Applications, IET 3(1), 1–19 (2009)
23. Yeol, J.W., Longman, R.W., Ryu, Y.S.: On the settling time in repetitive control systems. In: Proceedings of 17th International Federation of Automatic Control (IFAC) World Congress (July 2008)
24. Zaccarian, L., Teel, A.R.: A common framework for anti-windup, bumpless transfer and reliable designs. Automatica 38(10), 1735–1744 (2002)
25. Zheng, A., Kothare, M.V., Morari, M.: Anti-windup design for internal model control. International Journal of Control 60, 1015–1024 (1993)

Part III

Experimental Validation

6
Roto-Magnet

Summary. Systems with rotary elements are usually affected by periodic disturbances due to the movement of these parts (e.g. electrical machines, CD players). This kind of system is supposed to be moving, in some cases, at a fixed angular speed. Under these operating conditions any friction, unbalance or asymmetry appearing on the system generates a periodic disturbance that affects its dynamical behavior. In this context, RC has been shown as one of the most effective control strategies for these applications. Different test beds can be found in the literature which may be used to emulate pulsating periodic load or disturbance torques aimed at reproducing some particular real applications [4, 10]. In this chapter, an experimental plant intended for the educational purposes of analyzing and understanding the nature of the periodic disturbances as well as studying the different control techniques used to tackle this problem, has been adopted as experimental test bench for rotational machines.

The main goal of this chapter is to experimentally validate the proposals presented in previous chapters regarding RC working under time varying frequency conditions. In practice, varying period disturbances can be found in applications where the rotational speed may vary with time [2, 5]. Thus, aimed at providing varying-frequency disturbances, different reference profiles involving speed variations are employed.

The chapter is organized as follows. Section 6.1 presents the Roto-magnet plant, its concept, components and dynamic model. Section 6.2 describes the design of the standard RC for this plant. In Section 6.3 some implementation issues and two stability analysis for the RC working under varying sampling period are included. Section 6.4 and 6.5 are devoted to the implementation of the robust design method and the adaptive pre-compensation strategy, respectively. Finally, the optimal AW strategy is experimentally validated for HORC in Section 6.6.

6 Roto-Magnet

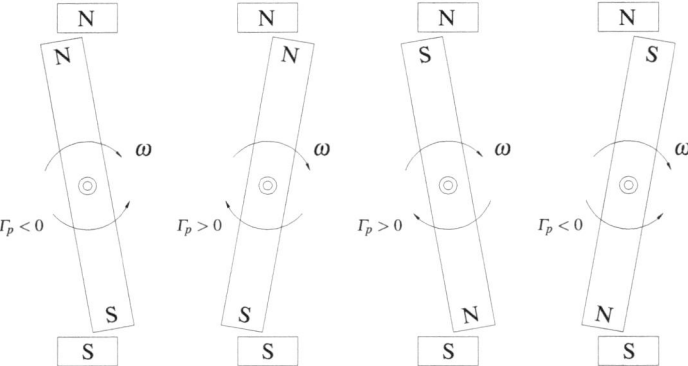

Fig. 6.1 Mechanical load: fixed and moving permanent magnets sketch (ω and Γ_p stand for the angular speed and the disturbance torque, respectively)

Fig. 6.2 Picture of the main part of the Roto-magnet plant: DC motor, optical encoder, magnetic system (load), and supporting structure

6.1 Plant Description

A picture of the Roto-magnet plant can be seen in Figure 6.2. It is composed of a bar holding a permanent magnet in each end, with each magnet magnetically oriented in the opposite way, and attached to a DC motor and two fixed electromagnets (see a sketch in Figure 6.1). The rotation of the DC motor causes a pulsating load torque (Γ_p) that depends on the mechanical angle θ of the motor axis. When the motor axis angular speed $\omega = \dot{\theta}$ is constant, i.e. $\dot{\omega} = 0$, the pulsating torque is a periodic signal with a fundamental period directly related to the axis speed: $T_p = \omega^{-1}$, with ω expressed in rev/s. The control goal for this plant is keeping the motor axis angular speed constant at a desired value.

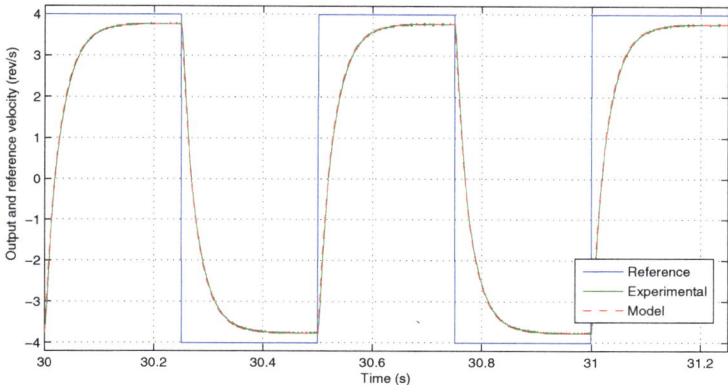

Fig. 6.3 Closed-loop time response of the system without the electromagnets

The plant system consists of the DC motor arranged with a power driver in current control loop configuration. Due to the high gain in open loop of this configuration the identification procedure is carried out in closed loop using a proportional gain controller [6]. In the present case a proportional gain equal to 1 was used. Thus, Figure 6.3 shows the closed-loop time response of the system without the electromagnets under a square input signal. The model used to approximate this closed-loop scheme is as follows:

$$G_{os}(s) = \frac{0.9417}{0.02665s+1} \frac{\text{rev/s}}{\text{V}}.$$

In this way the plant model can be inferred based on the knowledge of the controller [8]. The following plant model has been derived:

$$G_p(s) = \frac{K}{\tau s+1} = \frac{16.152}{0.457s+1} \frac{\text{rev/s}}{\text{V}}. \quad (6.1)$$

This is a first order parametrization with characteristic time response $\tau = 0.457\ s$ and stationary state gain $K = 16.152\ \text{rev}/(\text{V}\cdot\text{s})$.

6.2 Standard Repetitive Controller

The controller is constructed from (6.1), for a nominal speed of $\omega = 4\ \text{rev/s}$ and obtaining 250 samples per period, i.e. $N = 250$. These settings imply a nominal sampling period of $\bar{T} = T_p/N = 1/(\omega N) = 1$ ms. Under these assumptions the nominal discrete time plant is:

$$G_p(z) = G_p(z,\bar{T}) = \frac{K(1-e^{-\bar{T}/\tau})}{z - e^{-\bar{T}/\tau}} = \frac{0.0353}{z - 0.9978}. \quad (6.2)$$

70   6 Roto-Magnet

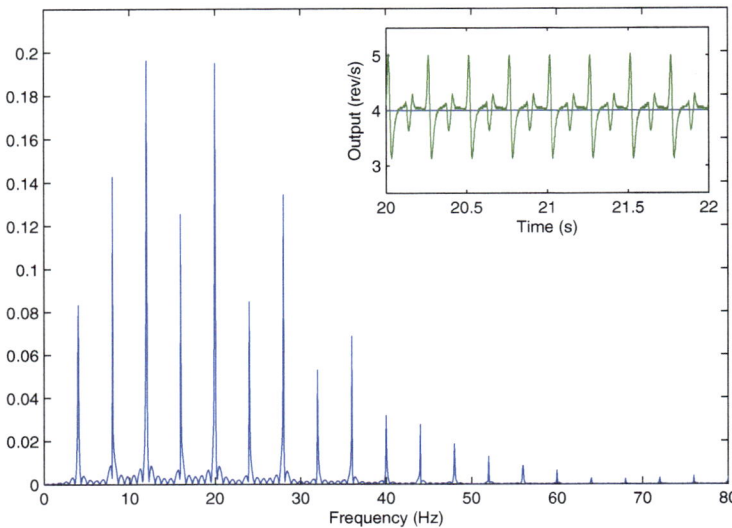

Fig. 6.4 Closed-loop time response and harmonic content without repetitive controller and with the electromagnets

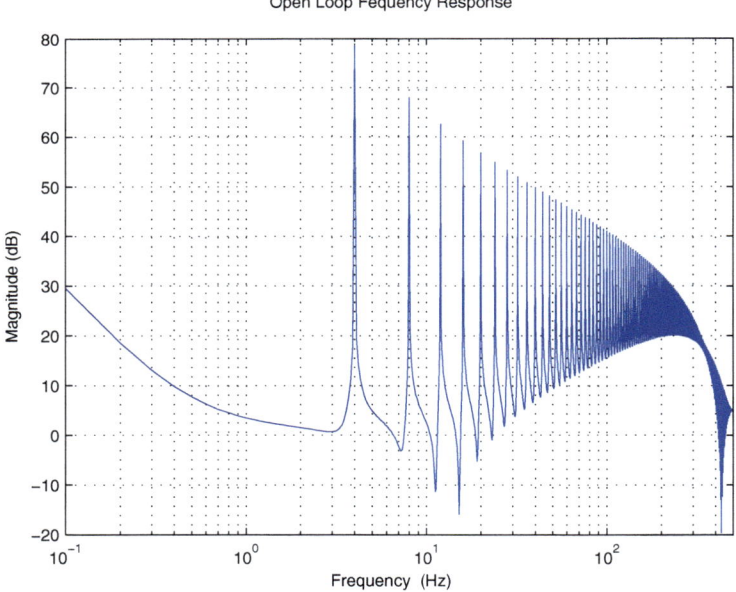

Fig. 6.5 Frequency response of the open-loop function $G_l(z)$

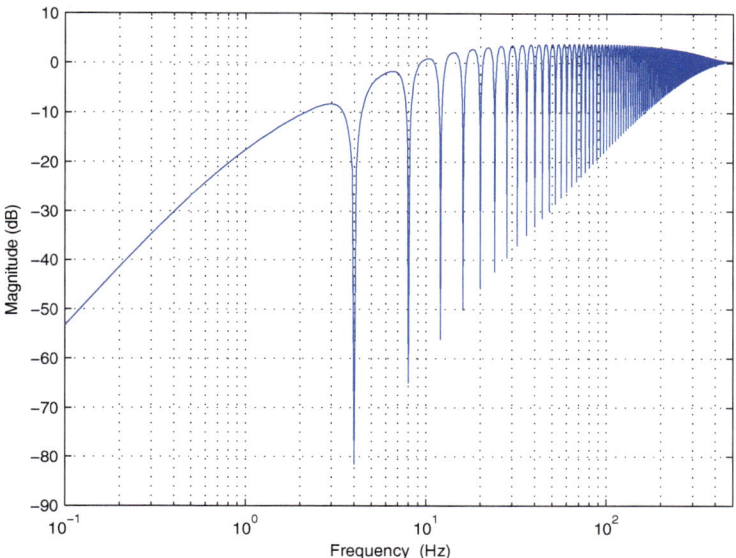

Fig. 6.6 Frequency response of the sensitivity function $S(z)$

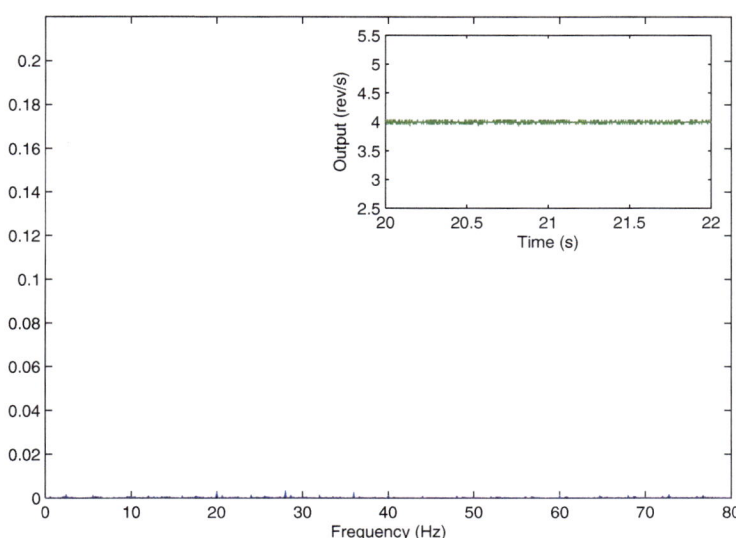

Fig. 6.7 Steady-state response and output signal harmonic content of the RC system

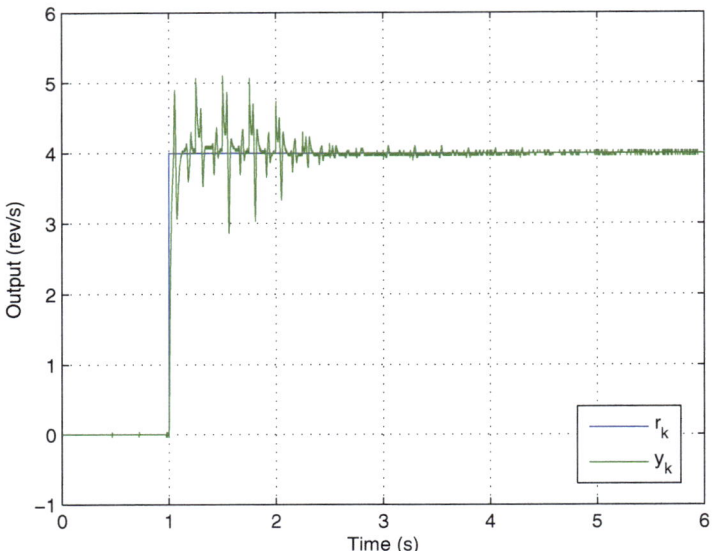

Fig. 6.8 Step response of the designed RC system

The standard full-harmonic IM will be used:

$$I_{st}(z) = \frac{H(z)}{z^N - H(z)}. \quad (6.3)$$

In accordance with Proposition 2.1 and the design procedure introduced in section 2.1.2, the following design issues have been taken into account:

- $G_c(z) = (1.8z - 1.796)/(z-1)$ provides a very robust inner loop.
- The first order null-phase FIR filter $H(z) = 0.25z + 0.5 + 0.25z^{-1}$ provides good performance in the present case.
- The fact that $G_p(z)$ is minimum-phase allows $G_x(z) = k_r G_o^{-1}(z)$, with $k_r = 0.7$.

These settings yield the control law:

$$\begin{aligned}
u_k &= 1.8e_k - 3.592e_{k-1} + 1.792e_{k-2} + 4.958e_{k-248} - 5.071e_{k-249} \\
&\quad - 9.916e_{k-250} + 10.14e_{k-251} + 4.958e_{k-252} - 5.07e_{k-253} \\
&\quad + 1.998u_{k-1} - 0.9978u_{k-2} + 0.25u_{k-249} + 0.0005556u_{k-250} \\
&\quad - 0.4994u_{k-251} - 0.0005556u_{k-252} + 0.2494u_{k-253},
\end{aligned} \quad (6.4)$$

with $e_k = r_k - y_k$, where y_k is the system output (speed) and r_k is the reference.

It is also useful to define the open-loop function $G_l(z) = (1 + I_{st}(z))G_c(z)G_p(z)$.

Figure 6.4 shows the closed-loop time response and the harmonic content[1] of the plant operating with the electromagnets, the controller $G_c(z)$ and without the

---
[1] Along this chapter the calculation of the harmonic content is carried out eliminating the constant component of the signals.

repetitive controller, i.e. the steady-state response of $G_o(z)$ with the disturbances. It is important to note that the speed describes an almost periodic signal. This type of disturbance may not be rejected by the controller $G_c(z)$.

The next experiment is performed with the repetitive controller. Figure 6.5 and 6.6 show the frequency response of the open-loop function $G_l(z)$ and closed-loop sensitivity function of the designed repetitive controller. These figures depict the action of the controller at the fundamental and harmonic frequencies. Figure 6.7 shows the experimental steady state output response at nominal frequency and its corresponding harmonic content. It can be seen that the system can successfully reject the periodic disturbances. Figure 6.8 depicts the transient response of the system after a step signal is applied as reference: the RC reject the disturbances after a short transient.

6.3 The Varying Sampling Period Strategy

6.3.1 Implementation Issues

The first issue to deal with in the implementation of the control action is the measurement of the uncertain or time-varying period T_p of the signal to be tracked or rejected. In the general case an adaptive scheme similar to those in [1, 5, 11] is used for this task; the complete controller architecture is depicted in Figure 3.1. The frequency values, $T_p^{-1}(t)$, are obtained by means of a frequency observer using information from different sources, namely, reference profile, output signal and control action. Then, the sampling rate is calculated as

$$T_s = \frac{T_p}{N}. \tag{6.5}$$

Nevertheless, as it has been previously stated, in mechanical turning systems the frequency to be tracked/rejected is directly related to the turning speed. Hence, for the analyzed mechatronic plant the disturbance frequency is straightforwardly computed from the turning speed reference. This allows us to study the effect of the adaptive approach on the closed-loop system decoupled from the frequency observer dynamics.

For the general case, the study of the effect of the frequency observer dynamics on the global system stability is out of the scope of this work. However, it is also worth recalling that any faulty estimation and/or implementation of T_s causes a difference between the experimental ratio T_p/T_s and the implemented value for N which, in turn, results in a performance degradation. As an example, Figure 6.9 shows the first harmonic gain factor evolution of the IM against a relative deviation of the sampling period T_s with respect to the nominal $\bar{T} = 1 \ ms$, namely

$$\frac{\left| I_{st}\left(\exp\left[j\frac{2\pi}{N(1+Q(T_s))\bar{T}}\bar{T}\right]\right) \right|}{\left| I_{st}\left(\exp\left[j\frac{2\pi}{N\bar{T}}\bar{T}\right]\right) \right|} = \frac{\left| I_{st}\left(e^{\frac{2\pi j}{N(1+Q(T_s))}}\right) \right|}{\left| I_{st}\left(e^{\frac{2\pi j}{N}}\right) \right|},$$

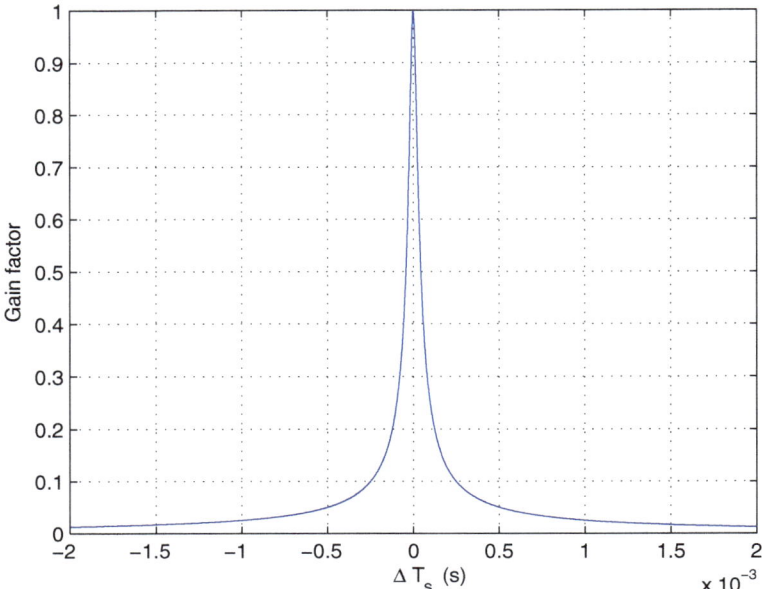

Fig. 6.9 First harmonic gain factor evolution at harmonic frequencies

where $I_{st}(\cdot)$ is defined in (6.3) and

$$Q(T_s) = \frac{T_s - \bar{T}}{\bar{T}}.$$

Notice that even small deviations of T_s entail important gain reductions. However, as stated in Remark 3.2.iv, closed-loop stability is not threatened unless $T_s \notin \mathscr{T}$.

Finally, and also in accordance with Remark 3.2.iv, the frequency observer output has to be saturated so as to guarantee that $T_s \in \mathscr{T}$.

Quantification Error Effect

It is well known that in a real setup T_s cannot be fixed with infinite precision but within the limitations imposed by the timer quantification. In the following, it will be shown that even small quantification errors entail important gain reductions.

The proposed implementation uses a digital counter as a time source, which offers a resolution of 1.6 μs. For a nominal sampling time of $\bar{T} = 1$ ms this implies a relative error of $1.6 \cdot 10^{-3}$, that may yield a gain reduction factor of 0.016 in the worst case, i.e. when approximating by truncation. Although the gain reduction is important, the sampling time adaptation scheme allows to maintain the gain of (6.3) above 40 dB, which is good enough for most applications.

6.3.2 LMI Gridding Approach

This Section is devoted to the stability analysis of the repetitive controller that employs the sampling adaptation mechanism in order to preserve the performance under varying frequency conditions.

The stability analysis of RC working with varying sampling period using the LMI gridding approach has been described in Section 3.3.

It is important to recall that, as mentioned in Remark 3.2, the high order involving the IM of the RC imposes a restriction in the applicability of the LMI approach. In this way, with the order of the controller obtained in the previous Section the resulting LMI problem is not computationally solvable. For that reason, in this Section a special setup of the RC for the mechatronic plant has been designed. Thus, the new setup is based on a nominal sampling period $\bar{T} = 5$ ms and nominal speed $\omega = 8$ rev/s which results in $N = 25$. This value of N is now appropriate to carry out the analysis. Also the following has been set:

- $G_p(z) = G_p(z, \bar{T}) = (0.1757)/(z - 0.9891)$.
- $G_c(z) = (1.8z - 1.78)/(z - 1)$.
- $H(z) = 0.02z + 0.96 + 0.02z^{-1}$.
- $G_x(z) = k_r G_o^{-1}(z)$, with $k_r = 0.7$.

Although the controller is designed to regulate the speed at 8 rev/s, in practice it will be necessary to move from this design point. Let us assume that we are interested in varying the speed reference in the interval $[6, 13]$ rev/s: this entails a sampling period variation in the interval $\mathscr{T} = [3.077, 6.667]$ ms.

To start the analysis procedure it is needed to write the closed-loop system in the form presented in Section 3.2, namely equation (3.2). Specifically, we need to find the matrix $\Phi(T_k)$ to derive the subsequent LMI problem.

The stability analysis that arises from Proposition 3.3 includes the solution of the LMI (3.6), which is known to be feasible, and the checking of the negative-definite character of $L_{T_k}(P_N)$. Figure 6.10 shows the evolution of the maximum eigenvalue of $L_{T_k}(P_N)$ when solving for $\alpha = 100$, and also for 50000 uniformly distributed values of T_k. Therefore, it can be presumed that the closed-loop system may operate in a speed range of $[7.32, 8.94]$ rev/s with dynamically preserved stability. This speed interval is obviously very narrow and operation conditions are limited to a sampling period interval $\mathscr{I}_N$ such that $\mathscr{T} \not\subset \mathscr{I}_N$. Once at this point, it is important to recall that this test comes not from a necessary condition but from a sufficient condition, so moving out of this interval does not necessarily imply instability[2].

In order to guarantee a broader stability interval, the second method described in Section 3.3 may be applied. Therefore, 40 uniformly distributed points are selected in $\mathscr{T} = [0.00307, 0.00667]$ s. These points are used to construct the set of LMIs (3.7), and a feasible solution $P = P_G$ with $\alpha = 100$ is obtained. Figure 6.11 depicts the maximum modulus eigenvalue of $L_{T_k}(P_G)$, detailing with a star the 40 points

---

[2] Although this step can be avoidable in most cases, its results are very useful to tune the LMI solver regarding next step.

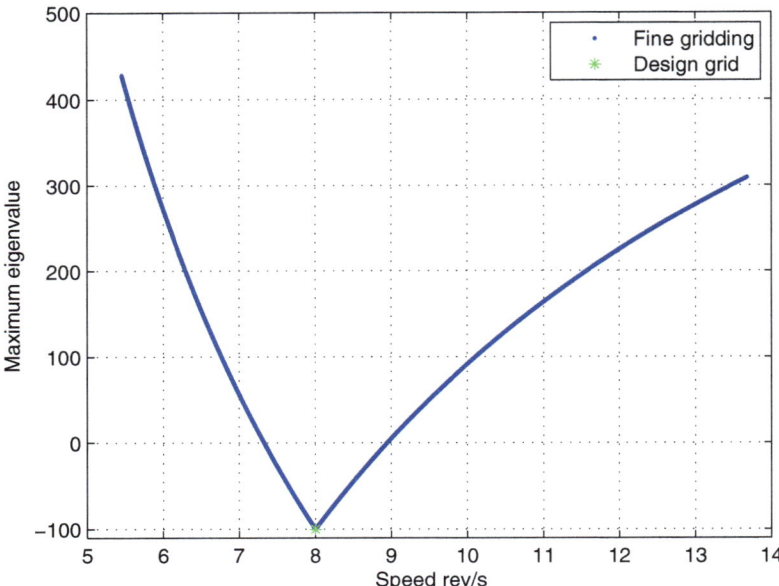

Fig. 6.10 Maximum eigenvalue of $L_{T_k}(P_N)$ with $\alpha = 4645$ and $\bar{T} = 0.005$s

leading to the LMI formulation. The maximum modulus eigenvalue of $L_{T_k}(P_G)$ corresponding to a finer grid consisting of 55121 uniformly distributed points is also drawn in Figure 6.11. These points are used to check the sign of $L_{T_k}(P_G)$ in the intervals between the points defining the LMI set. It can be seen that $L_{T_k}(P_G) < 0$ for every point in this finer grid of the interval $\mathscr{T}$; hence, stability is dynamically preserved therein. This method extends the previously obtained stability interval $[7.32, 8.94]$ rev/s, thus providing less conservative results. Further extensions of the new interval could also be feasible. It is also worth saying that the finer grid contains the points that correspond to the system clock resolution. Therefore, the stability of the digital implementation can be guaranteed in the obtained internal.

However, according to Remark 3.2, this LMI approach does not provide sufficient stability conditions and may entail numerical problems. Hence, the robust control approach results are given in next subsection.

6.3.3 Robust Control Theory Approach

The stability analysis method used in this section is based on robust control theory. The approach was previously described in Section 3.4 and the procedure renders sufficient conditions for the closed-loop stability of the RC system.

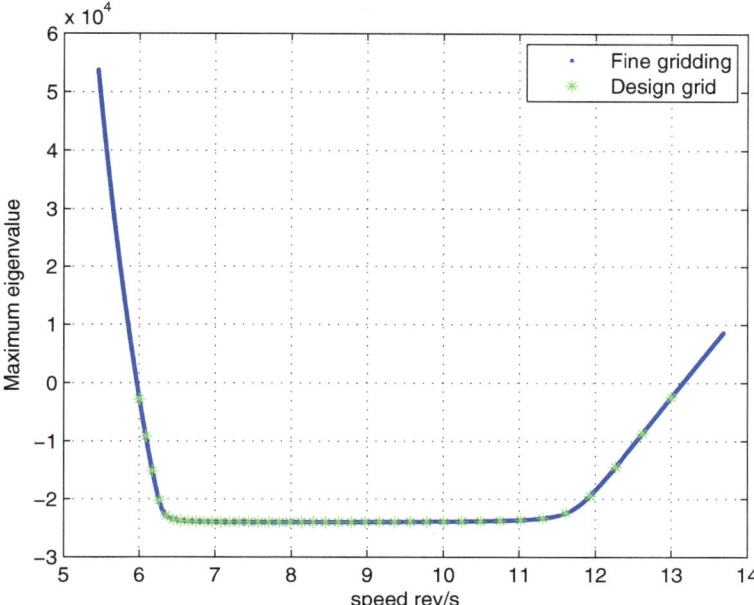

Fig. 6.11 Maximum eigenvalue of $L_{T_k}(P_G)$ with 40 points for the design grid and 55121 points to check stability

Firstly, one has to find matrix $\Phi(T_k)$ (equation (3.2)) and its representation in the form of Proposition 3.4 (recall equation (3.8)), and then re-write the system as in the system definition (3.15) and Figure 3.4.

Based on Theorem 3.1, in order to calculate the sampling period variation interval for which the system preserves the stability we have to calculate the H_∞-norm of system $G_{\bar{T}}(z)$.

Following Subsection 3.4, the settings for the plant and the repetitive controller in Section 6.2, i.e. $N = 250$ and $T_s = 1$ ms, yield $\|G_{\bar{T}}(z)\|_\infty = 1302.2$. In order to define $\gamma_{\bar{T}}$ (see (3.16)), $\varepsilon = 0.0001$ has been selected. Furthermore, the plant being first order yields a scalar value for its continuous-time system matrix: $A = -2.1876$; hence, an exact bounding of $\|\Delta(T_k - \bar{T})\|$ is possible. A straightforward calculation [3] shows that (3.17) is fulfilled with (see also Appendix B)

$$\mathcal{T} = \left[\bar{T} + \frac{1}{A}\log\left(1 - \frac{A}{\gamma_{\bar{T}}}\right), \bar{T} + \frac{1}{A}\log\left(1 + \frac{A}{\gamma_{\bar{T}}}\right)\right].$$

The specific numerical values are $\mathcal{T} = [0.2327, 1.7686]$ ms, which indicates that the performance of the device is ensured in the range $\omega \in [2.2617, 17.1906]$ rev/s.

With the settings specially applied for the LMI gridding approach (Section 6.3.2, i.e. $N = 50$ and $T_s = 5$ ms) the following has been found: $\|G_{\bar{T}}(z)\|_\infty = 273.5081$, $\mathcal{T} = [1.3583, 8.6709]$ ms and $\omega \in [4.6131, 29.4476]$ rev/s. These results show that the robust approach yields a wider stability interval compared with the LMI gridding approach.

6.3.4 Experimental Results

In order to illustrate the control system stability and performance, two different speed reference profiles have been used in the experimentation. Each profile corresponds with one of the two different settings described in the preceding section. The profile $\omega_{4ref}(t)$ and $\omega_{8ref}(t)$ for the $T_s = 1$ ms and $T_s = 5$ ms setup respectively:

$$\omega_{4ref}(t) = \begin{cases} 4 \text{ rev/s} & \text{if } t \in [0,15) \text{ s}, \\ -\frac{7}{32}t + \frac{233}{32} \text{ rev/s} & \text{if } t \in [15,19) \text{ s}, \\ 3.125 \text{ rev/s} & \text{if } t \in [19,28) \text{ s}, \\ \frac{25}{96}t - \frac{25}{6} \text{ rev/s} & \text{if } t \in [28,40) \text{ s}, \\ 6.25 \text{ rev/s} & \text{if } t \geq 45 \text{ s}. \end{cases}$$

$$\omega_{8ref}(t) = \begin{cases} 8 \text{ rev/s} & \text{if } t \in [0,15) \text{ s}, \\ -\frac{7}{16}t + \frac{233}{16} \text{ rev/s} & \text{if } t \in [15,19) \text{ s}, \\ 6.25 \text{ rev/s} & \text{if } t \in [19,28) \text{ s}, \\ \frac{25}{48}t - \frac{25}{3} \text{ rev/s} & \text{if } t \in [28,40) \text{ s}, \\ 12.5 \text{ rev/s} & \text{if } t \geq 45 \text{ s}. \end{cases}$$

During the time interval $[0,15]$ s, the reference is maintained constant at the nominal value $\bar{\omega}$. At $t = 15$ s a ramp reference change, from $\bar{\omega}$ to $0.7812\bar{\omega}$, is introduced in the system; then, the speed is kept constant for 9 s and finally at $t = 28$ s the speed is gradually augmented at a constant acceleration until it reaches the value $1.5625\bar{\omega}$ at $t = 40$ s.

Additionally, the experiment has been carried out for three different settings of the sampling time: keeping constant the sampling period at the nominal value $\bar{T}$, assuming a second order frequency observer for T_p and using an exact estimation of T_p. The following continuous model has been used to simulate the second order dynamics of the observer:

$$G_{obs}(s) = \frac{6.25}{s^2 + 2.5s + 6.25}.$$

This model has been implemented in discrete-time according to the selected sampling time.

In Figures 6.12.a and 6.13.a the sampling period is kept constant at the nominal value $\bar{T}$, for all t; it is important to realize that disturbances are almost rejected in $[0,15]$, that is, when both reference and sampling are at the nominal values. However, for $t > 15$ s disturbances cannot be properly compensated and performance is strongly degraded. Figures 6.12.b and 6.13.b depict the output behavior when the sampling period is varied adaptively assuming that the estimation of T_p uses a second order frequency observer; the profile of the actually used T_s are in Figures 6.14 and 6.15 respectively. Notice that the performance gets worse in the regions where

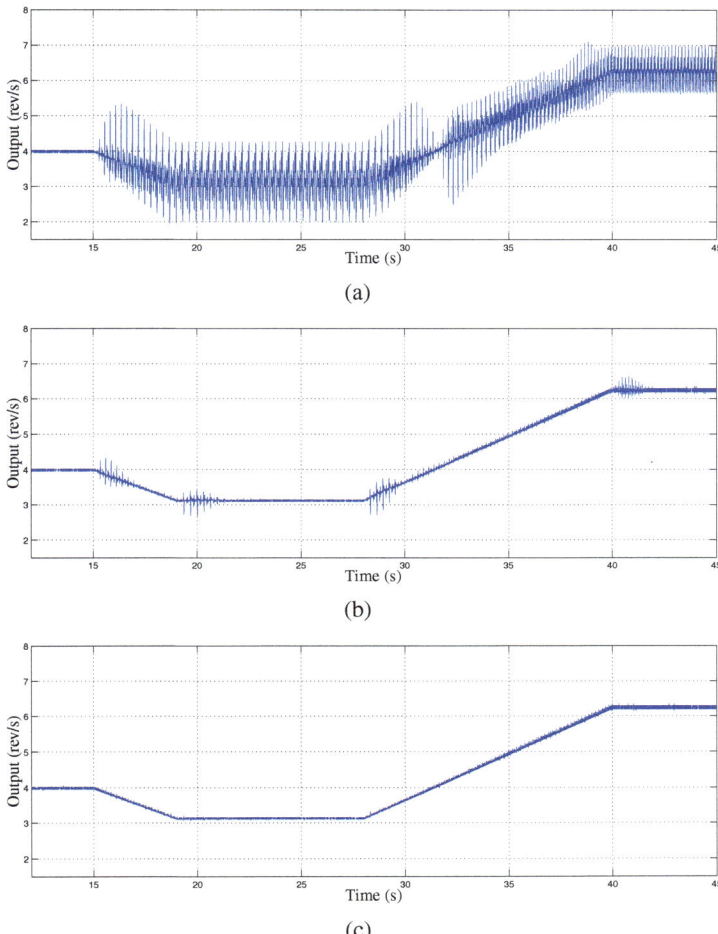

Fig. 6.12 Closed-loop system behavior using a repetitive controller and with sampling period T_s: (a) Fixed at the nominal value ($\bar{T} = 1$ ms); (b) Obtained from a second order frequency observer for T_p; (c) Obtained from an exact estimation of T_p

there is estimation error. Finally, Figures 6.12.c and 6.13.c portray the response under adaptive variation of T_s from an exact estimation of T_p. According to Sections 3 and 6.3.1, stability is preserved in the three situations because T_s always belongs to $\mathscr{T}$. Performance in the steady-state depends on the accuracy of the estimation of T_p, while performance during transients, although not guaranteed by RC theory, follows here the same pattern as that observed for the steady state, i.e. it improves with better estimations of T_p.

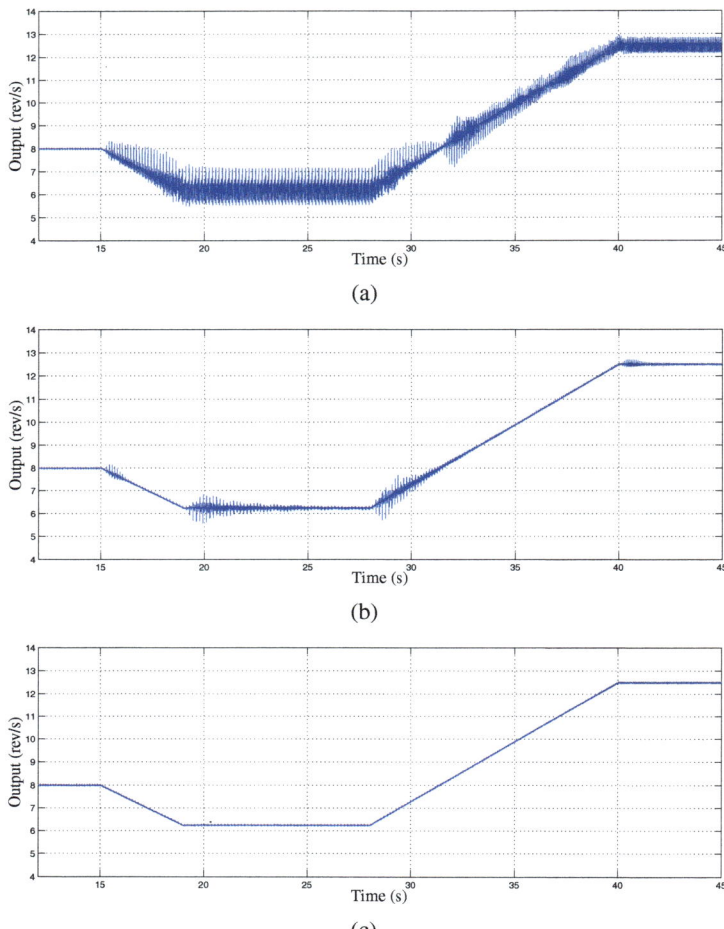

Fig. 6.13 Closed-loop system behavior using a repetitive controller and with sampling period T_s: (a) Fixed at the nominal value ($\bar{T} = 5$ ms); (b) Obtained from a second order frequency observer for T_p; (c) Obtained from an exact estimation of T_p

6.3 The Varying Sampling Period Strategy   81

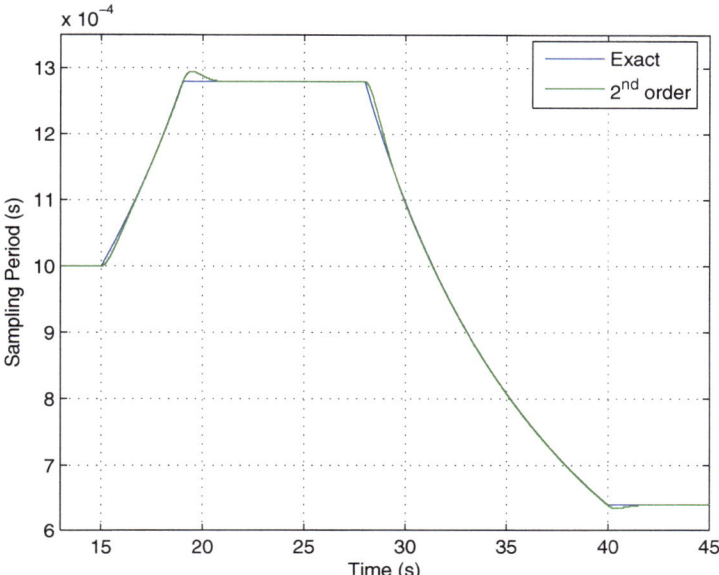

Fig. 6.14 Sampling period T_s corresponding to an exact (blue) and a second order (green) observer for T_p. Design for $\bar{T} = 1$ms.

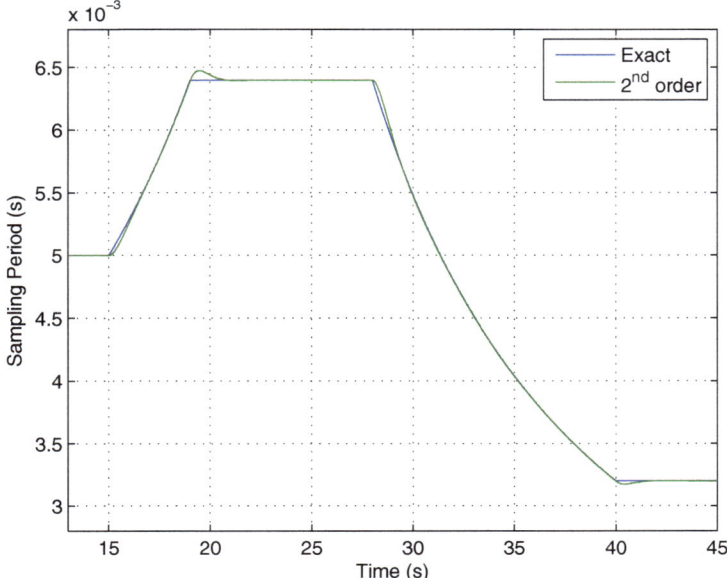

Fig. 6.15 Sampling period T_s corresponding to an exact (blue) and a second order (green) observer for T_p. Design for $\bar{T} = 5$ms.

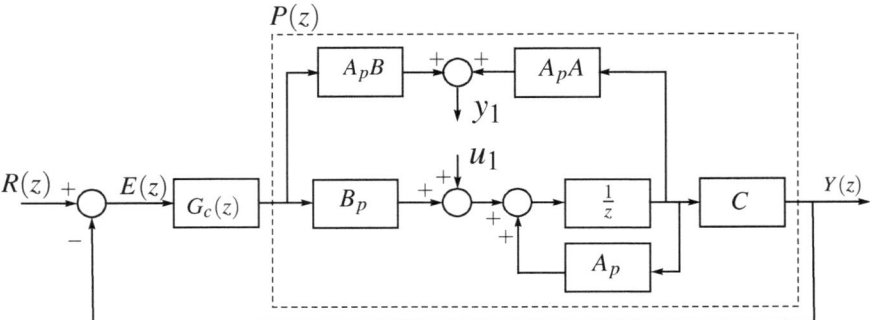

Fig. 6.16 Adaptation of the $G_o(z)$ control loop to the H_∞ formulation

6.4 Robust Design

The robust design proposal is a RC system design that takes into account the effect of the sampling period variations and yields a controller which assures stability for a given sampling period interval. In order to do that, the variation of the sampling period is rendered into a norm bounded uncertainty. The strategy is described in detail in Section 4.2.

Thus, following Section 4.2, the design may be performed in two stages: first, verify the design of $G_c(z)$ and second, re-design the stabilizing filter $G_x(z)$. To check that the controller $G_c(z)$ provides robust stability for the given sampling period variation interval one needs to fulfill the following condition (see Figures 6.16 and 6.17):

$$\|G^o_{T_s}(z)\|_\infty < (1+\varepsilon)^{-1} \|\Delta(T_s - \bar{T})\|_\infty^{-1},$$

which allows to calculate a sampling period interval $\mathscr{T}$ for which the system remains stable. For the experimental validation the design is based on an interval of $\omega \in [2.5, 8]$ rev/s. Thus, for the settings in Section 6.2, the following is obtained: $\|G^o_{\bar{T}}(z)\|_\infty = 1024.95$ and with this, the variation intervals results $\mathscr{T} = [0.0253, 1.9766]$ ms and $\omega \in [2.0235, 157.5851]$ rev/s. This means that with the selected $G_c(z)$ it is possible to perform a design for the speed variation interval of $\omega \in [2.5, 8]$ rev/s.

In the second stage, a μ-synthesis approach is used to take advantage from the problem structure. Figures 4.1 and 4.2 depict the scheme of the problem formulation and equation (4.2) shows the generalized plant needed to set up the problem. For the interval $T_s \in \mathscr{T} = \left[(2.5N)^{-1}, (8N)^{-1}\right]$, with $N = 250$, the uncertainty due to the sampling time variation can be estimated as: $\Delta_{T_s} \in [5.002, 5.996] \cdot 10^{-4}$. Thus, the procedure seeks a controller that, from the point of view of the delay block, provides a closed-loop system with H_∞ norm less than 1. With this information the μ-synthesis can be carried out and the following controller is obtained:

6.4 Robust Design

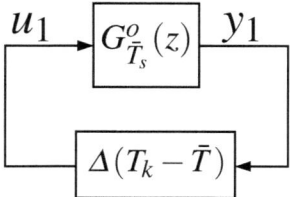

Fig. 6.17 The resulting feedback system with the uncertainty

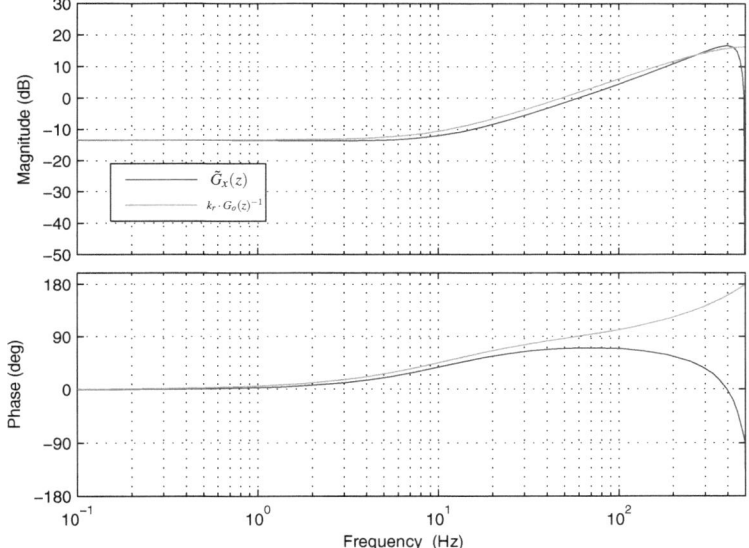

Fig. 6.18 Frequency response comparison between $\tilde{G}_x(z)$ and the original controller $G_x(z) = k_r G_o^{-1}(z)$, with $k_r = 0.21$

$$\begin{aligned}\tilde{G}_x(z) = &(2.908z^{16} - 26.6z^{15} + 104z^{14} - 218.3z^{13} + 236z^{12} - 36.29z^{11} \\ &-254.1z^{10} + 350.6z^9 - 187.7z^8 - 15.82z^7 + 84.18z^6 - 51.93z^5 \\ &+14.7z^4 - 1.683z^3 - 0.0006212z^2 + 2.565 \cdot 10^{-10}z + 9.445 \cdot 10^{-14}) \\ /&(z^{16} - 8.27z^{15} + 28.79z^{14} - 52.55z^{13} + 47.03z^{12} - 1.468z^{11} \\ &-40.55z^{10} + 36.96z^9 - 6.567z^8 - 9.728z^7 + 6.573z^6 - 0.9262z^5 \\ &-0.3955z^4 + 0.1141z^3 + 4.215 \cdot 10^{-15}z^2 + 1.294 \cdot 10^{-11}z + 4.761 \cdot 10^{-15}).\end{aligned}$$

Remark 6.1. The frequency response of $\tilde{G}_x(z)$ and the original controller $G_x(z) = k_r G_o^{-1}(z)$, with a new value of $k_r = 0.21$, is shown in Figure 6.18. It can be observed that the responses are very similar in the low and medium range. Therefore, for the purpose of keeping low the order of the controller, the structure $G_x(z) = k_r G_o^{-1}(z)$ with $k_r = 0.21$ has been implemented in the experimental setup.

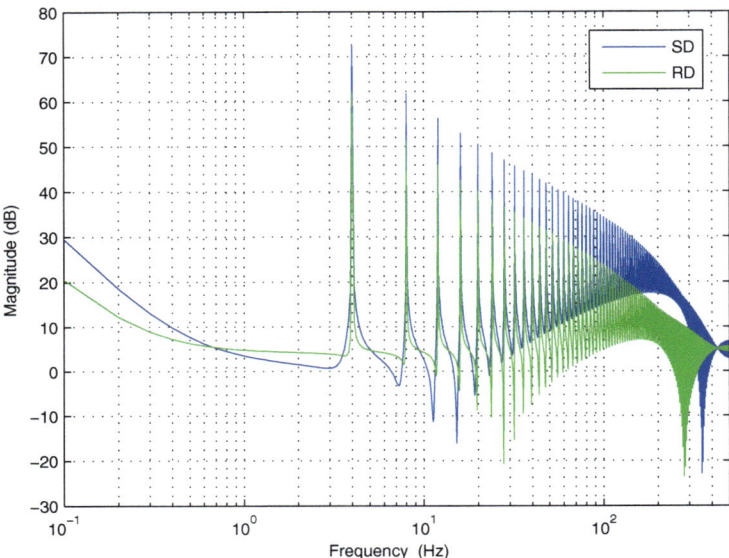

Fig. 6.19 Open-loop magnitude response of function $G_l(z)$. Standard design (SD) and robust design (RD) comparison.

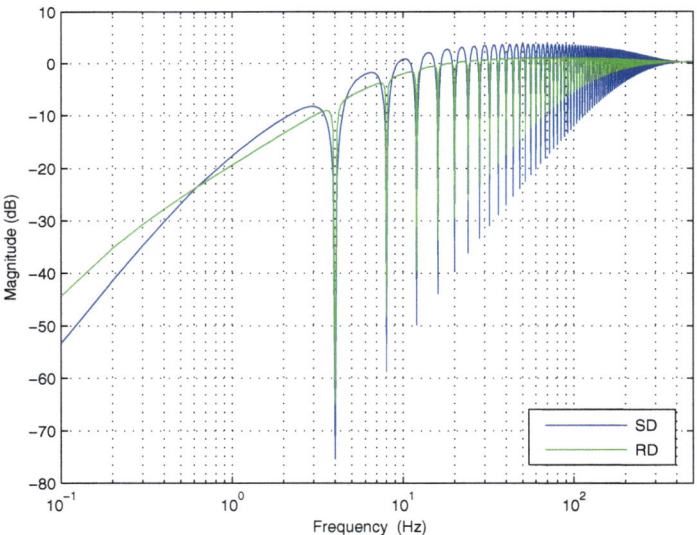

Fig. 6.20 Sensitivity function magnitude response. Standard design (SD) and robust design (RD) comparison.

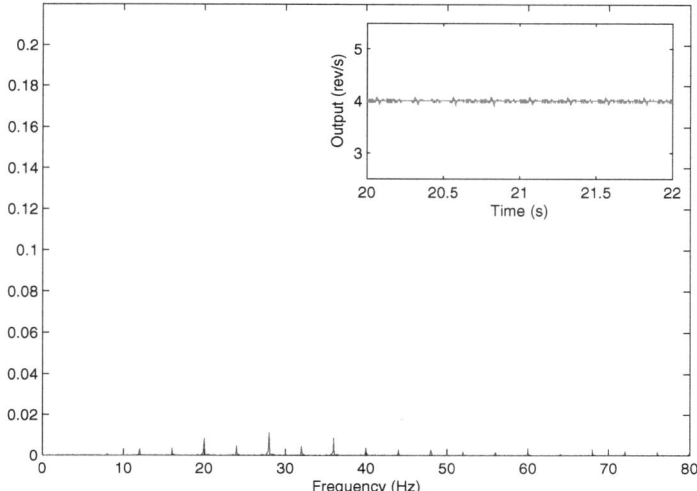

Fig. 6.21 Steady-state time-response and harmonic content using the RC robust design for nominal speed $\omega = 4$ rev/s

An important difference to be taken into consideration is that the standard design uses a stabilizing filter $G_x = k_r/G_o(z)$, which is a non causal transfer function, while the μ-synthesis design seeks a causal filter $\tilde{G}_x(z)$ to obtain a stable system for a given frequency variation interval.

Also it is valuable to note that, based on the IMP, the performance in steady state can be preserved if T_n, and in the same way T_s, remains constant for sufficiently large time intervals.

The open loop and sensitivity function frequency response of the RC designed in Section 6.2 and the robust design presented in this section using $\tilde{G}_x(z)$ are compared in Figures 6.19 and 6.20 respectively. It is shown that the tracking/rejection action is lower when using $\tilde{G}_x(z)$ since it corresponds to a smaller gain k_r. Additionally, in the sensitivity function comparison figure, it can be seen that the system using $\tilde{G}_x(z)$ exhibits better behavior at inter-harmonic frequencies.

6.4.1 Experimental Results

Figure 6.21 shows the steady-state time response and harmonic content of the RC robust design described in this section. Also, Figure 6.22 shows the output speed and control action evolution using the speed profile of Section 6.3.4 with $\bar{T} = 5$ ms and $N = 250$. In these figures it can be noticed that a small performance degradation appears due to the use of a smaller gain k_r; however the harmonic rejection is kept at a very acceptable level.

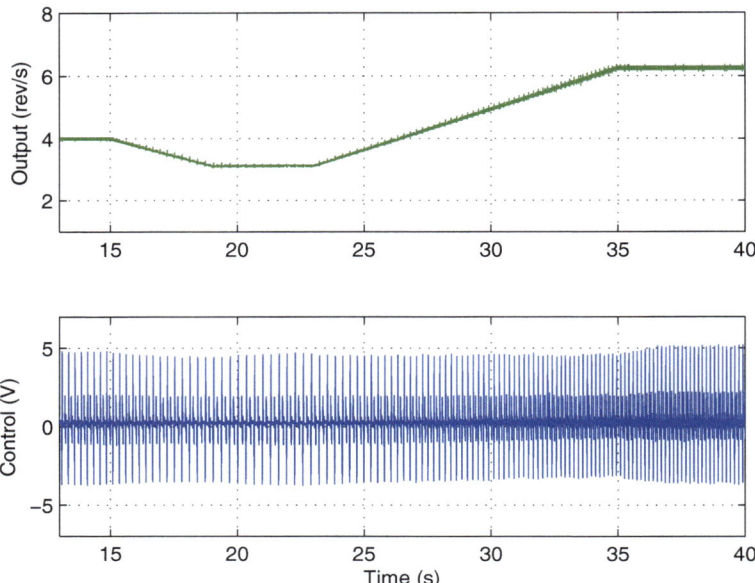

Fig. 6.22 Closed-loop system behaviour using the RC robust design with varying sampling rate T_s obtained from an exact estimation of T_p

6.5 Adaptive Pre-compensation

6.5.1 Controller Design

The aim of the adaptive pre-compensation design is to compensate for the structural variation effect encountered when using the sampling period adaptation mechanism. The proposal has been detailed in Section 4.3. According to that section, to force the system to behave as the one designed for the nominal sampling time the following pre-compensator needs to be included (see Figure 4.3):

$$C(z,T_s) = G_p(z,\bar{T})G_p^{-1}(z,T_s),$$

with

$$G_p(z,\bar{T}) \triangleq \frac{Num(z,\bar{T})}{Den(z,\bar{T})} = \mathscr{Z}\{G_p(s)\}$$

being the LTI model of the plant at nominal sampling time, and

$$G_p(z,T_s) = \frac{Num(z,T_s)}{Den(z,T_s)}$$

being the plant LTV model, i.e. working at varying sampling period.

For the Roto-magnet plant we have the following:

$$G_p(z,\bar{T}) = \frac{Num(z,\bar{T})}{Den(z,\bar{T})} = \frac{0.0353}{z - 0.9978} \tag{6.6}$$

and

$$G_p(z,T_s) = \frac{Num(z,T_s)}{Den(z,T_s)} = \frac{K(1 - e^{-T_s/\tau})}{z - e^{-T_s/\tau}}, \tag{6.7}$$

with K and τ set as in (6.1). Model (6.7) is first order, stable and minimum phase; hence, its inversion is possible and one can define:

$$\frac{Den(z,T_s)}{Num(z,T_s)} = \frac{z - e^{-T_s/\tau}}{K(1 - e^{-T_s/\tau})}. \tag{6.8}$$

Therefore, (6.6) and (6.8) yield

$$C(z,T_s) = \frac{0.0353 \left(z - e^{-T_s/\tau}\right)}{K(1 - e^{-T_s/\tau})(z - 0.9978)}, \tag{6.9}$$

which is a time-varying model that depends on the sampling period T_s and this, in turn, depends on the disturbance period variation T_p.

It is important to remark that, in this case, the function $P(z)$ introduced in (4.6) is composed of the series connection of:

1. The system $C(z,T_s)$ described in (6.9), which admits an LTV state-space representation $(A_c(k), B_c(k), C_c(k), D_c(k))$ with $A_c(k) = 0.9978$, $B_c(k) = 1$,

$$C_c(k) = \frac{0.0353(0.9978 - e^{-T_k/\tau})}{K(1 - e^{-T_k/\tau})},$$

$$D_c(k) = \frac{0.0353}{K(1 - e^{-T_k/\tau})},$$

the sampling period being $T_k = T_s(k) = t_{k+1} - t_k$.

Its internal stability is guaranteed by the fact that the state equation is time-invariant, with A_c being a 1×1 real matrix with modulus less than 1, which yields uniform exponential stability [7]. Moreover, as $B_c(k)$ is constant and $C_c(k), D_c(k)$ are bounded for all T_k belonging to any compact interval $\mathscr{T} \subset \mathbb{R}^+$, uniform BIBO stability is also guaranteed [7].

2. The system $G_p(z,T_s)$ described in (6.7), which is internally and uniformly BIBO stable because it corresponds to the sampled version of the LTI, continuous-time stable plant, $G_p(s)$.

Table 6.1 Normed error for the three proposed designs with time varying sample time: Standard, Pre-compensation and Robust

Error / RC Design	Standard	Pre-compensation	Robust
$\|e_k\|$	11.1996	17.2289	12.2048
$\|e_k\|_\infty$	0.1164	0.2088	0.1562

As the series connection of internally and BIBO stable systems is also internally and BIBO stable, in the present case $P(z)$ fulfills the hypothesis of Proposition 4.1, so the overall closed-loop system will be stable.

In order to derive the control action applied to the plant, i.e. the signal $\bar{U}(z)$ (see Figure 3.1), it has to be taken into account that the compensator makes the system time-invariant. Therefore, $\bar{u}_k$ is the invariant control law obtained from the nominal RC, that is:

$$\bar{u}_k = 1.8e_k - 3.592e_{k-1} + 1.792e_{k-2} + 4.958e_{k-248} - 5.071e_{k-249}$$
$$-9.916e_{k-250} + 10.14e_{k-251} + 4.958e_{k-252} - 5.07e_{k-253}$$
$$+1.998\bar{u}_{k-1} - 0.9978\bar{u}_{k-2} + 0.25\bar{u}_{k-249} + 0.0005556\bar{u}_{k-250}$$
$$-0.4994\bar{u}_{k-251} - 0.0005556\bar{u}_{k-252} + 0.2494\bar{u}_{k-253},$$

with $e_k = r_k - y_k$, r_k, y_k being, respectively, the system output and the reference velocities.

The derivation of u_k according to Figure 4.4 yields:

$$u_k = \frac{0.0353}{K(T_s)}\bar{u}_k - 0.0353\frac{1 - K(T_s)}{K(T_s)}\bar{u}_{k-1} + 0.9978u_{k-1},$$

with $K(T_s) = 1 - e^{-T_s/\tau}$. It follows by straightforward calculation than when the sampling interval remains constant at nominal the sampling time, i.e. $T_s = \bar{T}$, then $u_k = \bar{u}_k$.

6.5.2 Experimental Results

Figure 6.23 shows the experimental results using the adaptive pre-compensation scheme with varying sampling rate T_s, which is calculated using (6.5). One may observe that the controller can preserve the system performance. The control action, also shown in Figure 6.23, has a periodic behaviour and it can be seen how this period is changing so the controller can reject the varying frequency disturbances.

Additionally, Table 6.1 shows the error comparison for the three proposed designs that work with sampling time adaptation: the Standard, which corresponds with the conventional design, the Robust, that uses the robust control strategy, and the Pre-compensation, which lies on the pre-compensation technique. The results obtained in Figures 6.12, 6.22 and 6.23 with an exact estimation of T_p were used to calculate

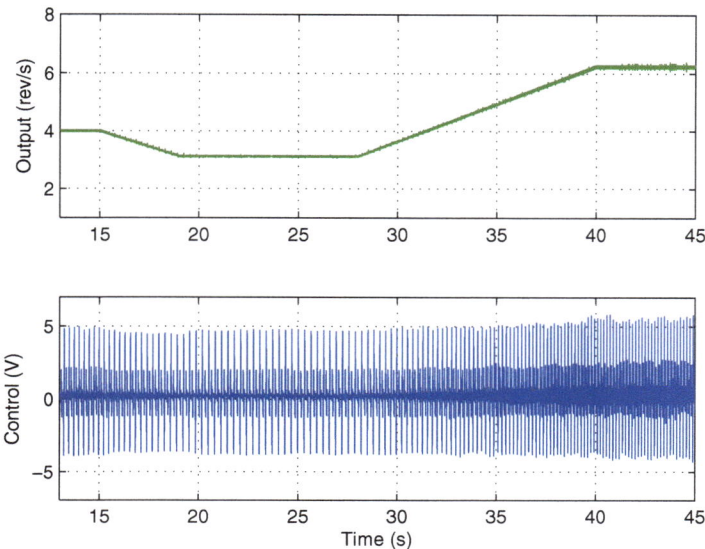

Fig. 6.23 Closed-loop system behaviour using a repetitive controller with adaptive pre-compensation with varying sampling rate T_s obtained from an exact estimation of T_p

the 2-norm and infinite norm of the resulting error. It can be noticed that the Standard design yields the smaller errors, followed by the Pre-compensation strategy, and finally, the Robust technique.

6.6 Anti-windup Optimal Design for HORC

In this section, an HORC is implemented on the Roto-magnet plant. The strategy is intended to provide robust performance against small speed variations which causes that the frequency of the disturbances varies in time. However, as mentioned in Section 5.2, the characteristics of the internal models used in HORC make the systems prone to the wind-up effect when there exists saturation in the control signal. For that reason it is necessary to include an AW compensator in the design. Thus, this section is aimed at experimentally validating the AW strategy described in Section 5.4 using the Roto-magnet plant. Two different tests have been developed: a disturbances rejection example at constant speed using the electromagnets of the plant and a periodic reference tracking example without the effect of the electromagnets.

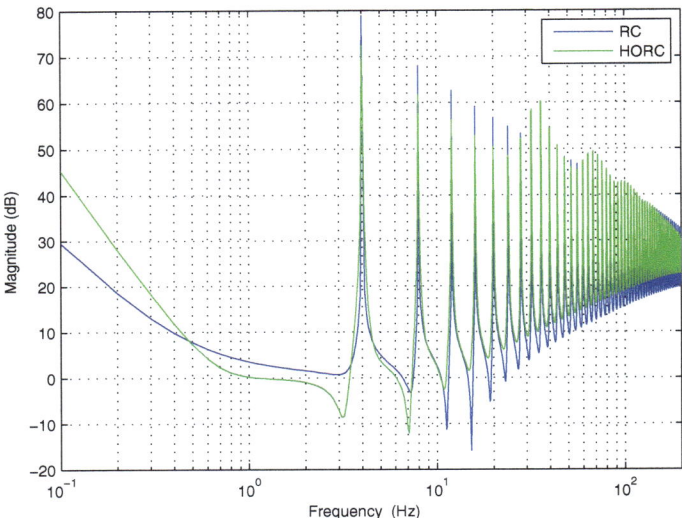

Fig. 6.24 Magnitude response of open-loop function $G_l(z)$. RC design and HORC design comparison.

6.6.1 Experimental Setup

The settings are the same used in Section 6.2 but changing the standard IM by a second order IM:

$$I_{ho2}(z) = \frac{\left(2z^{-N} - z^{-2N}\right) H(z)}{1 - \left(2z^{-N} - z^{-2N}\right) H(z)}$$

and defining

$$H(z) = 0.003297z^{-3} + 0.05897z^{-2} + 0.2492z^{-1} + 0.377 + 0.2492z + 0.05897z^2 + 0.003297z^3.$$

Figures 6.24 and 6.25 show the open-loop and sensitivity function magnitude response obtained with these settings. A comparison with the standard RC design of Section 6.2 has also been included. As it can be seen, the HORC offers a wider action around the harmonics compared with the standard RC. As discussed in previous chapters this effect provides the robustness against frequency variations. Figure 6.25 shows that HORC presents higher gain at inter-harmonic frequencies which in turn causes larger amplification at the high frequency zone. For that reason the bandwidth of the low-pass filter $H(z)$ has been reduced, thus providing the system with the appropriated stability robustness and, additionally, the gain $k_r = 0.7$ has been selected to reduce the inter-harmonic amplification (see Section 5.3.3).

The design of the AW compensator follows the procedure described in Section 5.4.6 and uses the structure of Figure 5.11. In this way, a compensator that has the dynamics of the plant during saturation and a deadbeat behaviour going out of saturation is designed.

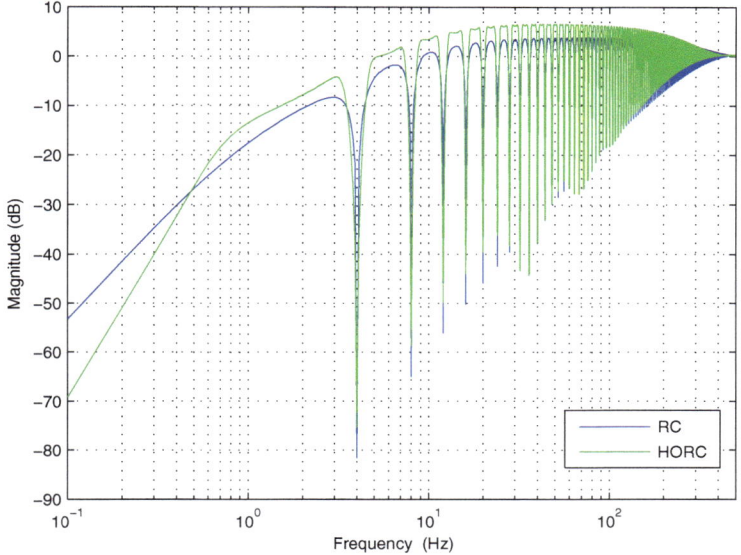

Fig. 6.25 Sensitivity function magnitude response. RC design and HORC design comparison.

Given the state space discrete-time model of the plant, (A,B,C,D), and its reachability matrix $W_A = \begin{bmatrix} B & AB & \cdots & A^{n-1}B \end{bmatrix}$, then $T = \begin{bmatrix} v_n^T & v_n^T A & \cdots & v_n^T A^{n-1} \end{bmatrix}$ with v_n^T being the last row of W_A^{-1}. Thus, for this example:

$$A = 0.9978, B = 0.15, C = 0.1412, D = 0, T = 4.$$

Using the optimal MRAW approach described in Section 5.4.6, the parameter that has been found to be a feasible solution is: $K_{db} = -3.991391$, using $Q_p = T^T T$ for a deadbeat approximation.

6.6.2 Experimental Results

The first experiment is aimed at showing the benefit of having an HORC to preserve performance in case of small variation/uncertainty of the disturbance period. Figures 6.26 and 6.27 depict the steady-state time response and harmonic content of the RC and HORC for a 0.5% deviation of the nominal speed $\omega = 4$ rev/s. It can be noticed that HORC obtains a lower degradation in the face of speed variation.

Anti-windup Experiments

The first experiment regarding the AW design problem is carried out in steady state for a nominal speed of $\omega = 4$ rev/s and with the disturbances generated by the electromagnets. The actuator saturation has been simulated limiting the control action at

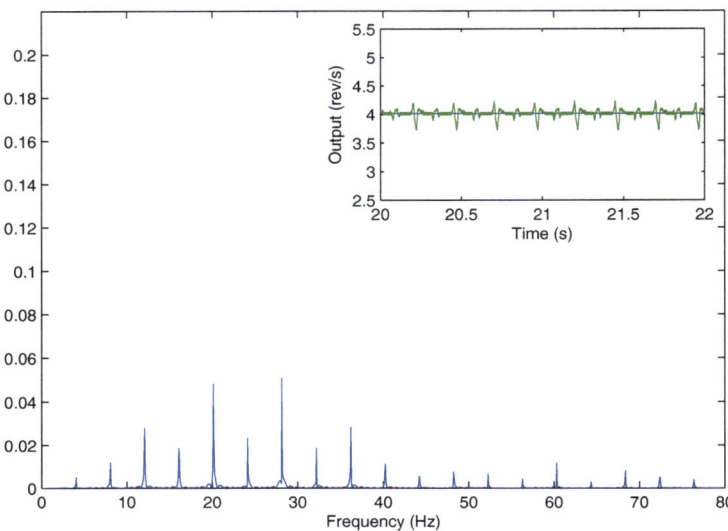

Fig. 6.26 Steady-state time-response and harmonic content of the RC for a 0.5% speed deviation

the output of the controller. Initially, the system has reached the steady state without saturation and at $t = 30$ s a limitation in the control action has been set at the value $u_{max} = 3$ V. Figures 6.28, 6.30 and 6.31 show the output signal and control action under three different cases: without any AW strategy, using the IMC AW technique and with the deadbeat optimal AW compensator proposed in this work, respectively. As it can be noticed, until $t = 30$ s, i.e. without saturation, the HORC can effectively reject the periodic disturbances generated by the electromagnets. However, without any AW strategy, Figure 6.28 shows that the system loses completely the performance and Figure 6.29 makes explicit the control action degradation after $t = 30$ s, where u_k starts growing, making it hard or impossible the recovery of the system. In the second case, using the IMC AW design, Figure 6.30 shows that even though the control action is not growing unbounded the system performance is very poor and it is in continuous degradation. In the third case, using the deadbeat AW optimal design, Figure 6.31 shows that although the system can not reject completely the disturbance due to the saturated control signal, the performance is almost completely recovered in the zones where the control action does not reach the saturation limit (see also Figure 6.32). Figure 6.32 shows the control action in detail: it can be noticed that the repetitive controller is providing the ideal control action and it is not affected by the saturation action while the plant receives the saturated version of this signal. Additionally, looking at the error signal, it is possible to see that the error signal grows only at time intervals when the control signal saturates.

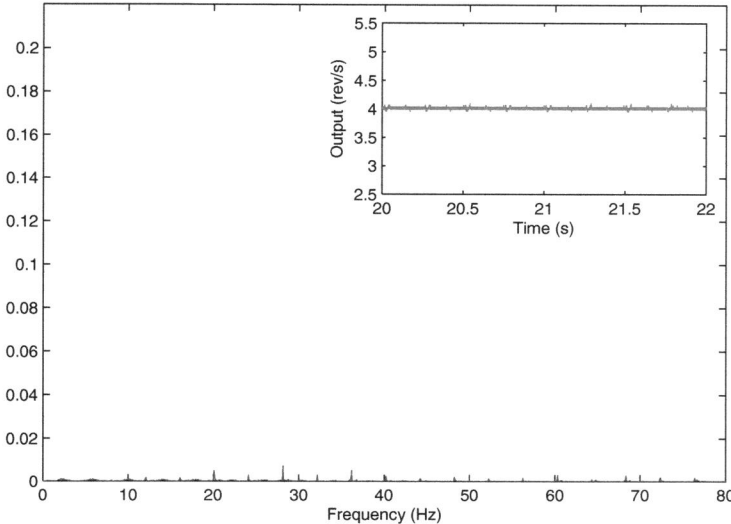

Fig. 6.27 Steady-state time-response and harmonic content of the HORC for a 0.5% speed deviation

The second experiment has been performed without the electromagnets and in this case it is a reference tracking example. Thus, the reference is a sinusoidal signal, and the saturation occurs limiting the control signal at $u_{max} = 3$. The same three cases analyzed in the previous experiment have been implemented. Figure 6.33, shows the output an control signal in steady state when there is no compensation for the saturation effect. It can be seen that the tracking performance can no longer be preserved and there is an important steady-state error. Figure 6.34 shows the detailed control action evolution and it is noticed that the controller output is reaching very high values making evident the windup effect. Figure 6.35 depicts the system response for the IMC AW design. It is shown that although the control action is bounded and the only difference with the ideal control signal is the truncation at the saturation limit, the tracking error is even larger than in the previous case since a constant deviation has been added. Figure 6.36 presents the system response when the deadbeat optimal design is applied. It is remarkable that the tracking error only appears when the system gets into saturation and it vanishes as soon as the system can reach the reference signal again. The control signal is similar to the previous one but in this case the system remains saturated longer allowing a faster recovery of the tracking performance for the rest of the period. The error signal $e_k = r_k - y_k$ of the three strategies is shown in Figure 6.37. It is evident that the optimal AW design exhibits the smallest error for this RC system.

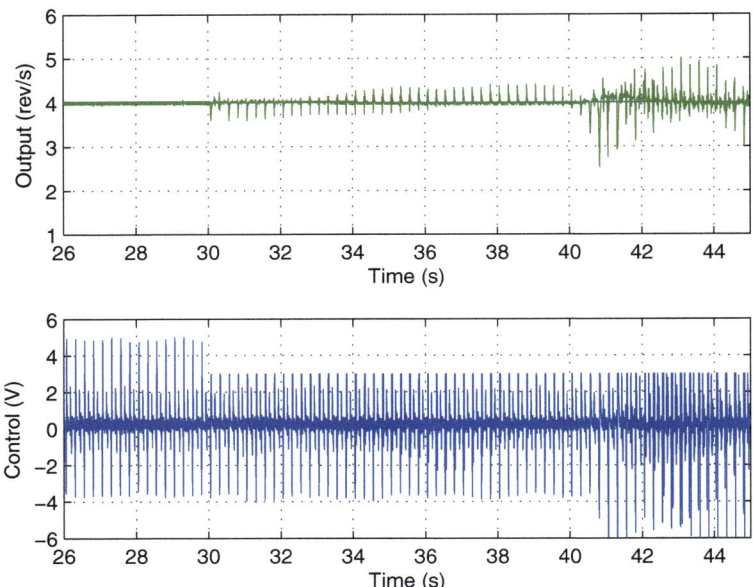

Fig. 6.28 Steady-state time-response of the system without AW compensator

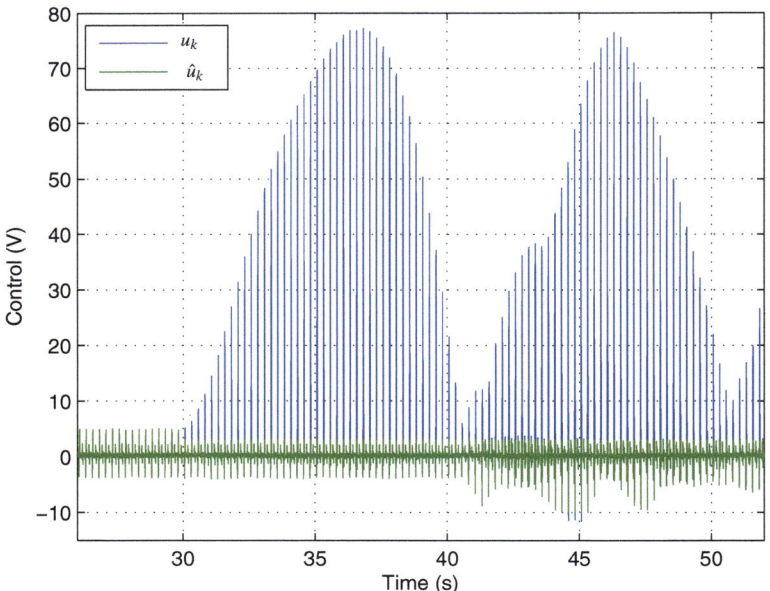

Fig. 6.29 Detailed control action of the system without AW compensator

6.6 Anti-windup Optimal Design for HORC 95

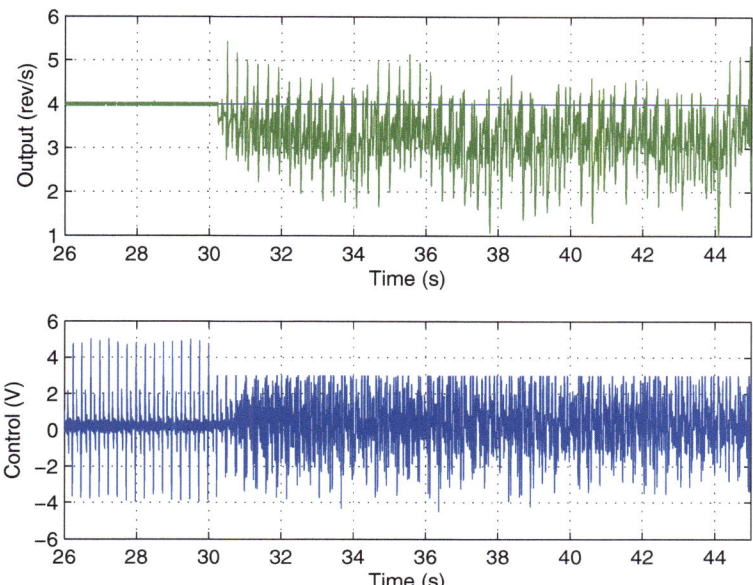

Fig. 6.30 Steady-state time-response of the system with IMC AW compensator

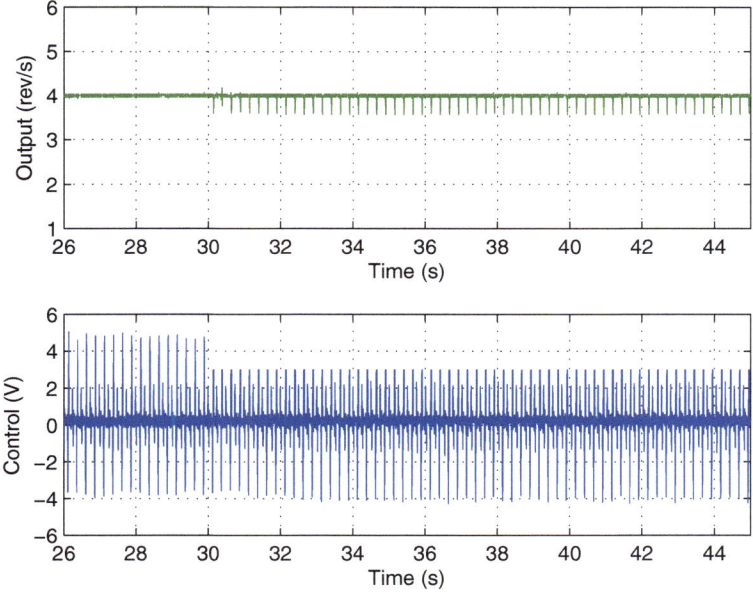

Fig. 6.31 Steady-state time-response of the system with deadbeat optimal AW compensator

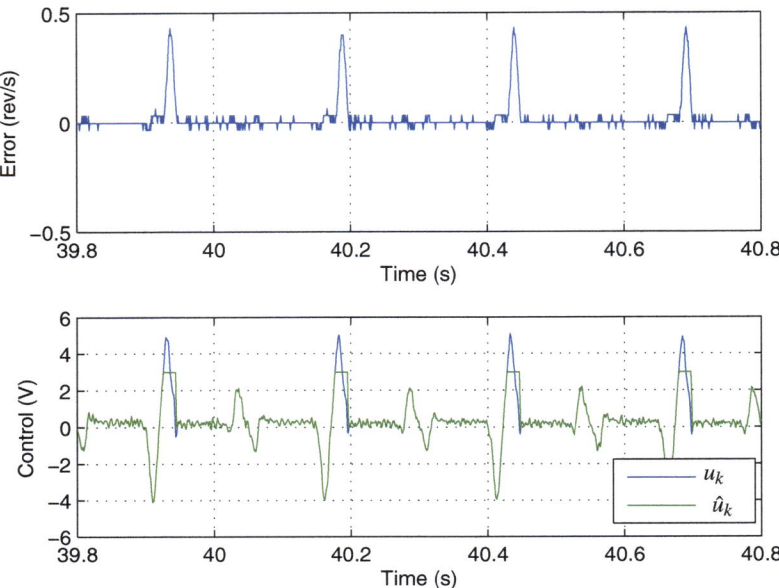

Fig. 6.32 Detailed error $e_k = r_k - y_k$ and control signal of the system with deadbeat optimal AW compensator

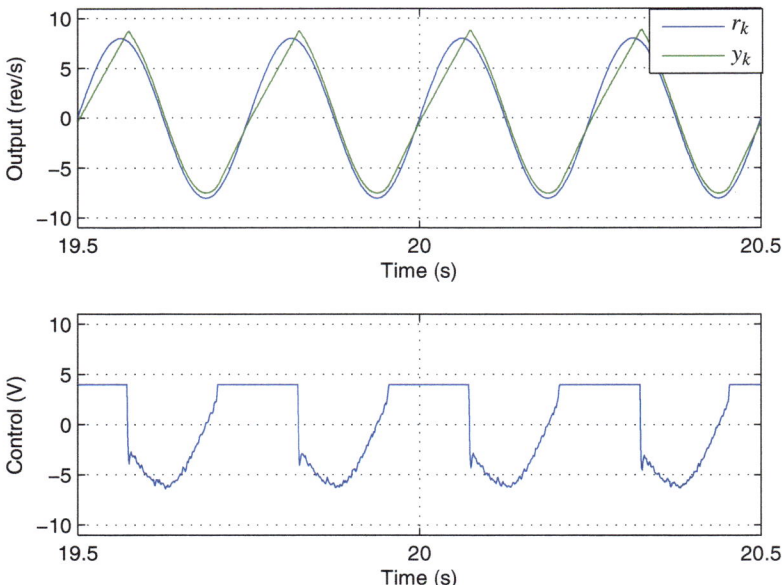

Fig. 6.33 Steady-state time-response of the system without AW compensator

6.6 Anti-windup Optimal Design for HORC 97

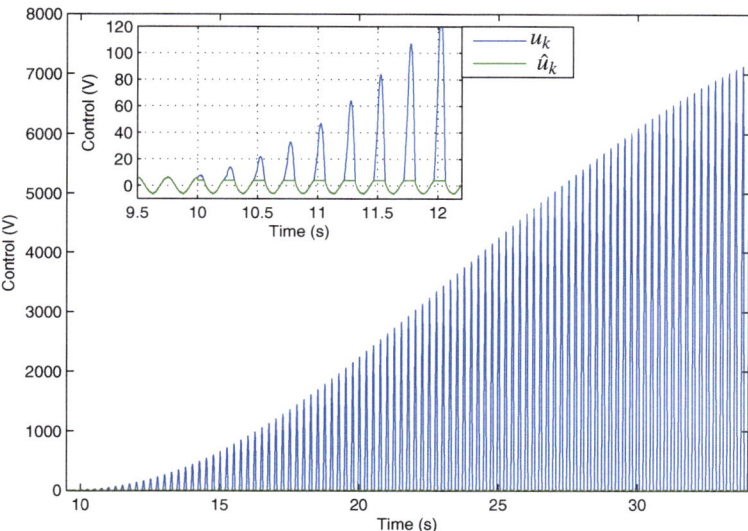

Fig. 6.34 Detailed control action of the system without AW compensator

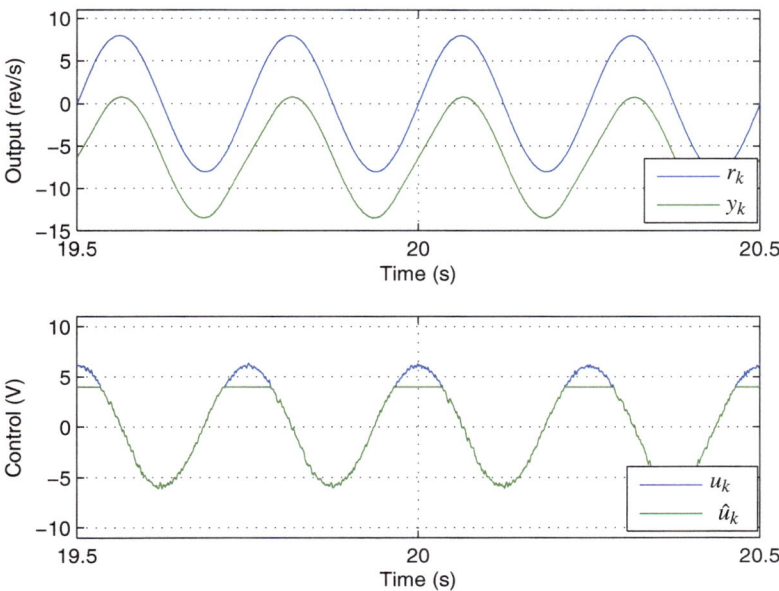

Fig. 6.35 Steady-state time-response of the system with IMC AW compensator

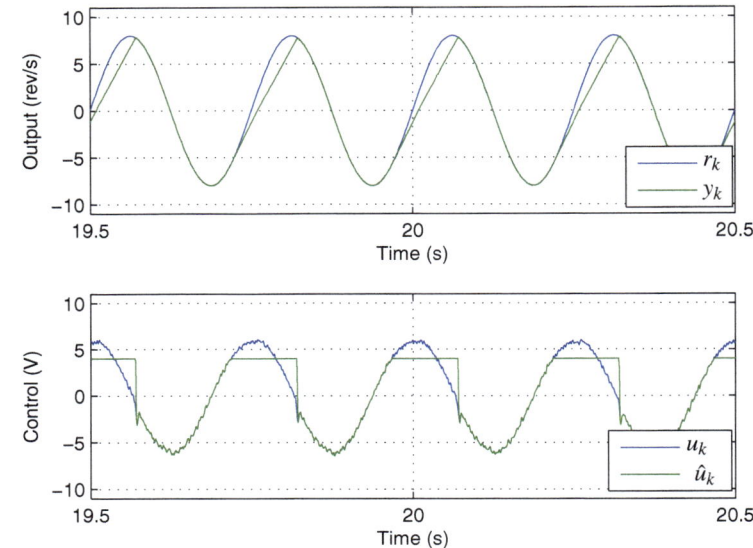

Fig. 6.36 Steady-state time-response of the system with DB optimal AW compensator

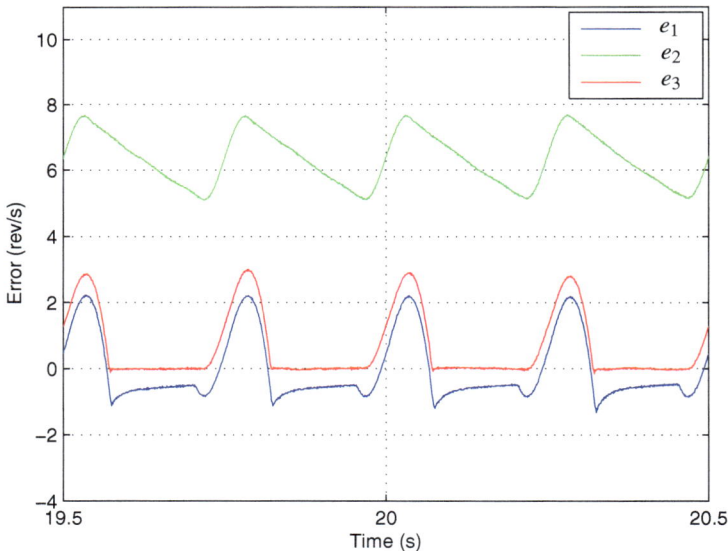

Fig. 6.37 Steady-state error comparison of the three AW strategies. e_1 stands for no AW compensation, e_2 for IMC AW strategy and e_3 for the optimal AW proposal.

6.7 Conclusions

Section 6.3 analyses the BIBO stability of a closed loop system containing a digital repetitive controller working under time-varying sampling period. The analysis was carried out using LMI gridding and robust control theory approaches that allow a stability assessment in a known, bounded interval where the reference/disturbance period is assumed to vary. The theoretically predicted results have been experimentally validated through the Roto-magnet plant in which rejection of periodic disturbances with time-varying period is successfully achieved. A detailed description of fundamental issues related to the implementation procedure have been provided.

The pre-compensation scheme proposed in Section 4.3 was implemented in Section 6.5. Thus, in addition to the sampling period adaptation mechanism a pre-compensator is included in the system in order to eliminate the structural changes due to the varying sampling period. As a consequence, closed-loop stability can be assured despite the sampling period adaptation and the frequency observer dynamics.

Section 6.4 used the robust control theory to synthesise a controller that ensures system stability for a predefined sampling period variation interval. As a result, it was found that the original controller $G_x(z)$ designed using the traditional methods is very closely related to the controller calculated using this strategy. This is consistent with the idea that the traditional design may be close to the optimum [9].

In Section 6.6, the optimal LQ design of Section 5.4 was synthesized for the Roto-magnet plant aimed at finding a deadbeat recovery behaviour and assuring the global asymptotic stability of the closed loop system. Through experimental results it is shown that the proposed AW scheme gets better performance when compared with other designs with the same structure, thus providing the smallest deviation from the ideal plant output and isolating the repetitive controller from the saturation effect.

Future research should study the inclusion of less-restrictive sector conditions for the nonlinear saturation function so as to better approximate the deadbeat design.

References

1. Cao, Z., Ledwich, G.F.: Adaptive repetitive control to track variable periodic signals with fixed sampling rate. IEEE/ASME Transactions on Mechatronics 7(3), 374–384 (2002)
2. Dotsch, H., Smakman, H., Van den Hof, P., Steinbuch, M.: Adaptive repetitive control of a compact disc mechanism. In: Proceedings of the 34th IEEE Conference on Decision and Control, vol. 2, pp. 1720–1725 (December 1995)
3. Fujioka, H.: A discrete-time approach to stability analysis of systems with aperiodic sample-and-hold devices. IEEE Transactions on Automatic Control 54(10), 2440–2445 (2009)
4. Grcar, B., Cafuta, P., Stumberger, G., Stankovic, A.: Control-based reduction of pulsating torque for pmac machines. IEEE Transactions on Energy Conversion 17(2), 169–175 (2002)

5. Hanson, R.D., Tsao, T.-C.: Periodic sampling interval repetitive control and its application to variable spindle speed noncircular turning process. Journal of Dynamic Systems, Measurement, and Control 122(3), 560–566 (2000)
6. Landau, I.D., Zito, G.: Digital Control Systems Design, Identification and Implementation. Springer (2006)
7. Rugh, W.: Linear system theory, 2nd edn. Prentice-Hall, Inc., Upper Saddle River (1996)
8. Söderström, T., Stoica, P.: System identification. Prentice-Hall, Inc., Upper Saddle River (1988)
9. Songschon, S., Longman, R.W.: Comparison of the stability boundary and the frequency response stability condition in learning and repetitive control. International Journal of Applied Mathematics and Computer Science 13(2), 169–177 (2003)
10. Su, T., Hattori, S., Ishida, M., Hori, T.: Suppression control method for torque vibration of ac motor utilizing repetitive controller with fourier transform. IEEE Transactions on Industry Applications 38(5), 1316–1325 (2002)
11. Tsao, T.-C., Qian, Y.-X., Nemani, M.: Repetitive control for asymptotic tracking of periodic signals with an unknown period. Journal of Dynamic Systems, Measurement, and Control 122(2), 364–369 (2000)

7
Shunt Active Power Filter

Summary. Shunt active power filters have been introduced as a way to overcome the power quality problems caused by nonlinear and reactive loads [9, 14, 23]. These power electronics devices are designed with the goal of obtaining a power factor close to 1 and achieving current harmonics and reactive power compensation [5, 6, 15]. The usual approaches [5, 15] for the control of shunt active filters are based on two hierarchical control loops: an inner one that assures the desired current and an outer one in charge of determining its required shape and the appropriate power balance as well. The control structure followed in this Chapter is the one in [7], in which the current controller is composed of a feedforward action that provides very fast transient response, and also of a feedback loop which includes an odd-harmonic repetitive control that yields closed-loop stability and a very good harmonic correction performance. In turn, the outer control law is based on the appropriated computation of the amplitude of the sinusoidal current network and, aiming at a robustness improvement, this is combined with a feedback control law including an analytically tuned PI controller.

However, although the control system performance is very good, it shows a dramatic performance decay when the network frequency value is not accurately known or changes in time. For a better assessment of this issue this Chapter shows the experimental behaviors under constant and varying network frequency. This performance degradation is also presented and analyzed in terms of the THD[1], PF[2] and $cos\phi$.

The Chapter organization is as follows. Section 7.1 introduces the plant, the control objectives and the two hierarchical control loops. Section 7.2 shows the odd harmonic controller designed for constant network frequency, also analysing the performance degradation through the THD, PF and $cos\phi$. Section 7.3 details the results for the varying sampling time repetitive controller including the stability

---

[1] The THD is calculated using $THD = \sqrt{\Sigma_{l=2}^{p} u_l^2}/u_1$, where u and l stand for the signal magnitude and harmonic order respectively.

[2] The PF calculation can be done with $PF = cos\phi/\sqrt{1+THD^2}$, where $cos\phi$ is the difference in phase between the voltage signal and the fundamental component of the current signal at the source.

analysis by means of robust control theory. It is worth to say that the stability analysis using the LMI approach introduced in Section 3.3 can not be applied in this implementation since the size of the resulting matrices makes the problem computationally unsolvable. Section 7.4 and Section 7.5 present the implementation of the adaptive pre-compensation scheme and the robust design respectively and, finally, a second order HORC is applied in Section 7.6.

7.1 Plant Description

Figure 7.1 shows the system configuration. The active filter is connected in parallel between the power source and the load aiming at guaranteeing unity PF at the network side. The active filter is composed of a boost converter with the ac neutral line connected directly to the middle point of the dc bus. The dynamic behavior of the boost converter can be expressed using the model averaged at the switching frequency as follows:

$$L\frac{di_f}{dt} = -r_L i_f - v_1 \frac{d+1}{2} - v_2 \frac{d-1}{2} + v_n, \tag{7.1}$$

$$C_1 \frac{dv_1}{dt} = -\frac{v_1}{r_{C,1}} + i_f \frac{d+1}{2}, \tag{7.2}$$

$$C_2 \frac{dv_2}{dt} = -\frac{v_2}{r_{C,2}} + i_f \frac{d-1}{2}, \tag{7.3}$$

where d is the control variable; i_f is the inductor current; v_1, v_2 are the dc capacitor voltages, respectively; $v_n = V_n\sqrt{2}\sin(\omega_n t)$ is the voltage source, with $\omega_n = 2\pi/T_n$,

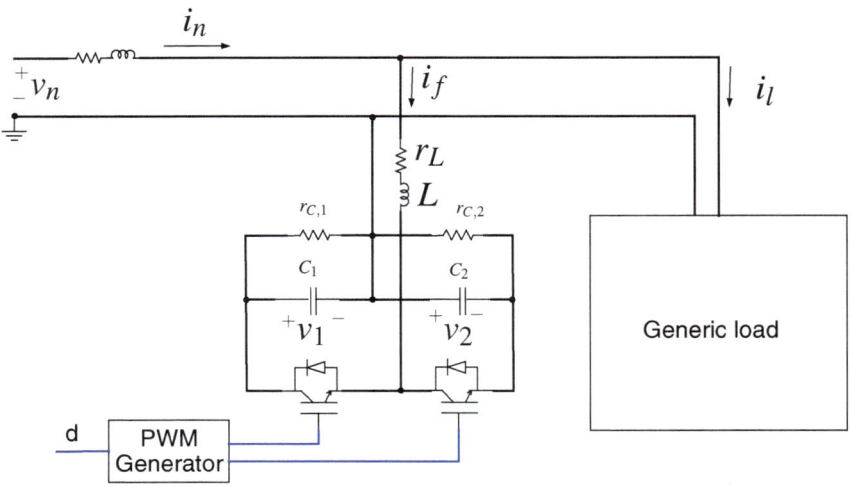

Fig. 7.1 Single-phase shunt active filter connected to the network-load system

T_n being the network period; L is the converter inductor; r_L is the inductor parasitic resistance; C_1, C_2 are the converter capacitors and $r_{C,1}, r_{C,2}$ are the parasitic resistances of the capacitors. The control variable, d, takes its value in the closed real interval $[-1, 1]$ and it is directly related to the PWM control signal injected to the actual system.

Because of the nature of the voltage source, the steady-state load current is usually an odd-symmetric periodic signal. Hence, its Fourier series expansion can be written as

$$i_l(t) = \sum_{n=0}^{\infty} a_n \sin\left[(2n+1)\omega_n t\right] + b_n \cos\left[(2n+1)\omega_n t\right].$$

Let us consider the partial state feedback $\alpha = \frac{d+1}{2}v_1 + \frac{d-1}{2}v_2$ and the variables

$$E_C = \frac{1}{2}\left(C_1 v_1^2 + C_2 v_2^2\right), \quad D = C_1 v_1 - C_2 v_2,$$

where E_C is the energy stored in the converter capacitors and D is the charge unbalance between them. Let us also assume that the two dc bus capacitors are equal ($C = C_1 = C_2$, $r_C = r_{C,1} = r_{C,2}$). Then, the system dynamics on the new variables is

$$L\frac{di_f}{dt} = -r_L i_f + v_n - \alpha \tag{7.4}$$

$$\frac{dE_C}{dt} = -\frac{2E_C}{r_C C} + i_f \alpha \tag{7.5}$$

$$\frac{dD}{dt} = -\frac{1}{r_C C}D + i_f. \tag{7.6}$$

7.1.1 Control Objectives

The objective of the active filter is to ensure that the current at the source, $i_n(t)$, has a sinusoidal shape in phase with the network voltage profile. This can be stated as[3] $i_n^* = I_d^* \sin(\omega_n t)$. In this way, it is required that the control system tracks a sinusoidal shape signal with suitable magnitude and rejects the harmonics introduced by the load and any existing phase difference.

Additionally, for the correct operation of the boost converter, it is necessary to assure a constant average value of the dc bus voltage[4], namely, $< v_1 + v_2 >_0^* = v_d$, where v_d must fulfil the boost condition ($v_d > 2\sqrt{2}v_n$). It is also desirable for this voltage to be almost equally distributed among both capacitors ($v_1 \approx v_2$). This can be reached as a consequence of seeking the active power balance of the system, which is achieved if the energy stored in the active filter capacitors, $E_C = v_1^2 + v_2^2$, is equal to a reference value, E_C^d.

---

[3] x^* represents the steady-state value of signal $x(t)$.
[4] $<x>_0$ means the dc value, or mean value, of the signal $x(t)$.

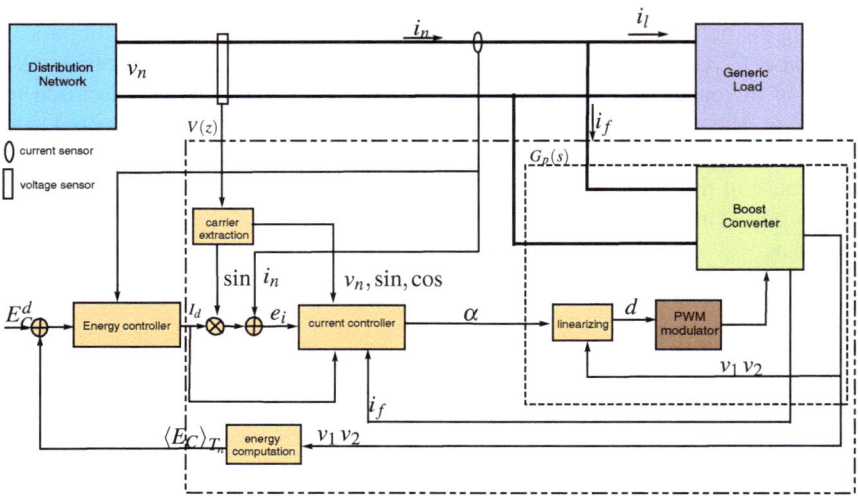

Fig. 7.2 Global architecture of the control system

7.1.2 Controller Structure

The controller proposed in this work is based on the hierarchical approach studied in [7] and shown in Figure 7.2. Firstly, an inner current controller is used to force the sine wave shape for the network current by fixing α and, and, consequently, d. Then, the appropriate active power balance for the whole system is achieved by an outer energy shaping control loop that sets the amplitude of the sinusoidal reference, I_d, for the current control loop. It is also worth remarking that a carrier extraction filter (see Figure 7.2) cleans the source voltage from possible background harmonic distortion.

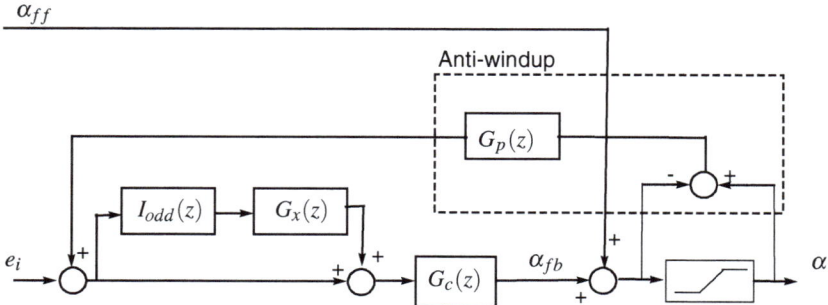

Fig. 7.3 Current control block diagram

The Current Loop Controller

The current control loop, depicted in Figure 7.3, aims at forcing a sinusoidal shape in i_n. Notice that its output signal, i.e. the control action α, is composed of a feedforward term, α_{ff}, and feedback term, α_{fb} [7].

The feedforward term is computed through the analysis of the control action required to obtain the desired steady state behaviour [7], and it is also in charge of providing fast transient response:

$$\alpha_{ff} = v_n + L\frac{di_l}{dt} + r_L i_l - I_d \left(r_L \sin \omega_n t + L\omega_n \cos \omega_n t \right). \tag{7.7}$$

The feedback term, which is introduced to overcome model uncertainties, disturbances and measurement noise, is designed taking advantage of the linearity of (7.4). This controller uses a repetitive control under a plug-in scheme. Thus, the implementation of the approaches proposed in previous chapters will be described in the subsequent sections.

As usual, the active filter control action is subject to saturation. This nonlinearity may introduce a windup effect [12] in the repetitive controller that should be avoided. However, the high order inherent to repetitive controllers limits the applicable AW techniques [11]. Specific AW schemes for RC have been proposed in [16, 18], but most of them increase the necessary computational burden. This work uses a strategy which allows a simple implementation: it consists in placing the plant model in the AW feedback chain as shown in Figure 7.3. This scheme coincides with the IMC AW approach described in Section 5.4.2 which may be viewed as a particular case of the MRAW scheme [11]. Hence, the real plant sees the saturated control action while the plant model sees the difference between the saturated and the unsaturated control, which makes the system LTI from the controller side. Consequently, the AW feedback loop stability is proved by construction. It is worth to say that we skip the optimal AW design proposed in Section 5.4.6 with the purpose of maintaining the active filter control design simpler.

The control action is the PWM duty cycle, d, which in practice belongs to the interval $[-0.975, 0.975]$. Hence, the AW scheme, as shown in Figure 7.3, is applied over α in the interval $[0.025v_1 - 0.975v_2, 0.975v_1 - 0.025v_2]$.

The Energy Shaping Controller

Following [7], once $i_n(t) \approx I_d(t)\sin(\omega_n t)$ is achieved, the outer plant can be modelled as:

$$\frac{d}{dt}\langle E_C(t)\rangle_{T_n} = \frac{-2}{Cr_C}\langle E_C(t)\rangle_{T_n} + \frac{\sqrt{2}V_n\pi}{\omega_n}(I_d - a_0). \tag{7.8}$$

The control of this plant is carried out through a two-action controller [7], as pictured in Figure 7.4. The first is a feedforward action term, defined as $I_d^{ff} = a_0$, which assures the energy balance in the ideal case ($r_L = 0$, $r_C = 0$) and takes into account the features and changes of i_l instantaneously. The second action is given by the PI controller:

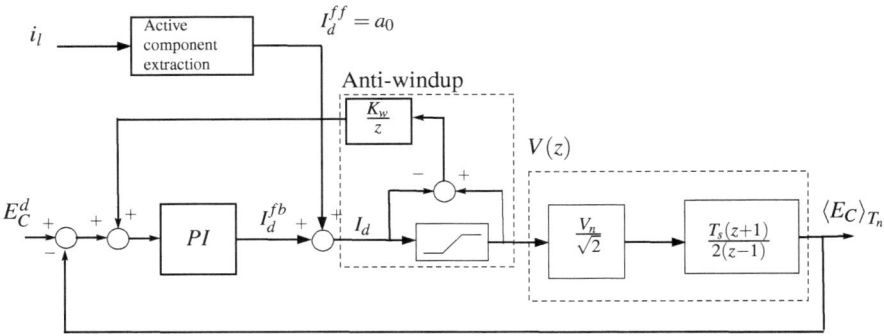

Fig. 7.4 Energy shaping controller block diagram

$$I_d^{fb} = k_i \frac{T_s(z+1)}{2(z-1)} \Delta E + k_p \Delta E, \qquad (7.9)$$

where $\Delta E = E_C^d - \langle E_C(t) \rangle_{T_n}$, which regulates $\langle E_C(t) \rangle_{T_n}$ to the set point E_C^d with zero steady-state error.

The implementation of this controller requires the computation of the mean value of certain signals. In this work the digital notch filter [4]

$$F_N(z) = \frac{1}{N} \cdot \frac{1 - z^{-N}}{1 - z^{-1}} \qquad (7.10)$$

has been utilized to compute $\langle \cdot \rangle_{T_n}$. It is important to emphasize that the notch filter used the fact that $E_C(t)$ is a periodic signal and annihilates all its periodic components.

Finally, the PI controller is completed with a standard AW mechanism that bounds the desired active power which the active filter may handle (see Figure 7.4). For the current setup it is $I_d \in [-40, 40]\,\text{A}$.

7.2 Odd Harmonic Repetitive Controller

The nominal period of the signal to be tracked/rejected is $T_p = 1/50$ s, which corresponds with the nominal European network frequency of 50 Hz. The sampling period is selected to be $T_s = 5 \cdot 10^{-5}$s (the PWM switching period), this yielding $N = T_p/T_s = 400$.

The discrete-time plant model is obtained from (7.1) and the addition of an anti-aliasing filter with time constant τ:

$$G_p(z) = \left(1 - z^{-1}\right) \mathscr{Z} \left[\frac{-1}{Ls + r_L} \cdot \frac{1}{\tau s + 1} \cdot \frac{1 - e^{-T_s}}{s} \right]_{T_s}, \qquad (7.11)$$

which gives a minimum-phase system. The inductor and the parasitic resistance values are $L = 1$ mH and $r_L = 0.5\,\Omega$ respectively. For the inner loop we use the lag controller

$$G_c(z) = -\frac{3.152z - 3.145}{z - 0.9985},$$

which provides a phase margin of 79.36°. The odd-harmonic IM (5.5) is selected:

$$I_{odd}(z) = \frac{-H(z)}{z^{\frac{N}{2}} + H(z)}, \qquad (7.12)$$

with

$$H(z) = \frac{1}{4}z + \frac{1}{2} + \frac{1}{4}z^{-1},$$

while

$$G_x(z) = k_r(G_o(z))^{-1}, \qquad (7.13)$$

and $k_r = 0.7$. The open-loop function is defined as

$$G_l(z) = (1 + I(z))G_c(z)G_p(z),$$

with $I(z)$ corresponding to the IM.

The repetitive controller yields the feedback law

$$\alpha_{fb} = G_c(z)\left[1 + G_x(z)I_{odd}(z)\right]e_i$$

which, together with the feedforward action given in (7.7), yields the control action $\alpha = \alpha_{fb} + \alpha_{ff}$. Under its combined effect, it can be assured that the network current is $i_n(t) \approx I_d(t)\sin\omega_n t$, which will be taken as a fact.

The frequency response of the open loop and sensitivity function are depicted in Figures 7.5 and 7.6, respectively. These Figures show the tracking and rejection action of the repetitive controller over the harmonic frequencies.

7.2.1 Performance at Nominal Frequency

Figure 7.7 shows the distribution network voltage, v_n, and the network current, i_n, when a rectifier (i.e. a nonlinear load) is connected to the network and no active filter is in use[5]. This current has a THD[6] of 62.6% and a RMS value of 19.56 A, while the PF amounts to 0.76. The function of active filters is to inject the current required to transform the network current i_n into a sinusoidal one with low THD in v_n and i_n.

Figure 7.8 shows the experimental system response at nominal frequency. As it can be seen, the control system can effectively reject the harmonic components present in the load current achieving a sinusoidal shape current at the source, i_n, with a THD of 0.3% with unitary PF and $\cos\phi$. Additionally, Figure 7.9 depicts the transition registered in i_n when the load is switched on. As shown, the system reaches the steady state in approximately 7 cycles.

---

[5] The scales in the Figures of this chapter are: v_n (230 V/div), i_n (48 A/div) and v_1, v_2 (74,5 V/div).

[6] In this work, the THD is calculated with respect to the RMS value of the signal.

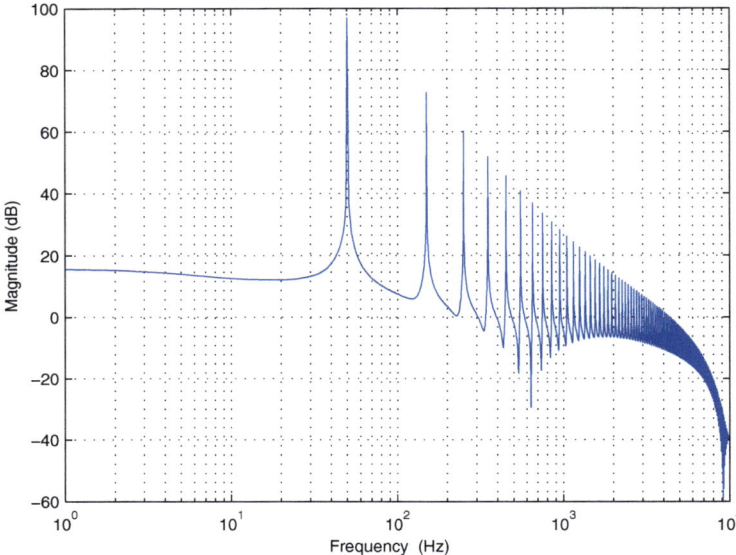

Fig. 7.5 Magnitude response of the open-loop function $G_l(z)$ for the active filter RC design

The next experimental setup consists of a linear load composed of a capacitive-resistive array, the previously described single phase active filter and second order odd-harmonic controller. Figure 7.10 shows the waveforms of v_n and i_n when the linear load is connected to the ac source. The system PF and $cos\phi$ are 0.75.

As Figure 7.10 shows, when the frequency of the voltage source is fixed to 50Hz and the active filter is connected in parallel with the linear load, the PF and $\cos\phi$ at the port are unitary.

7.2 Odd Harmonic Repetitive Controller

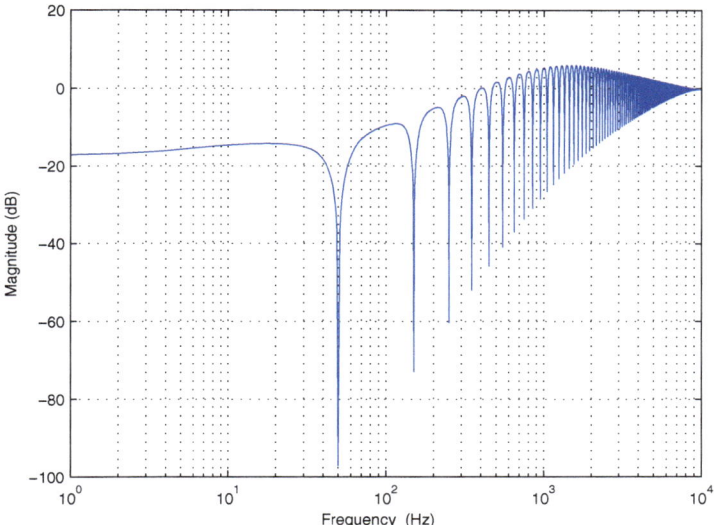

Fig. 7.6 Magnitude response of the sensitivity function for the active filter RC design

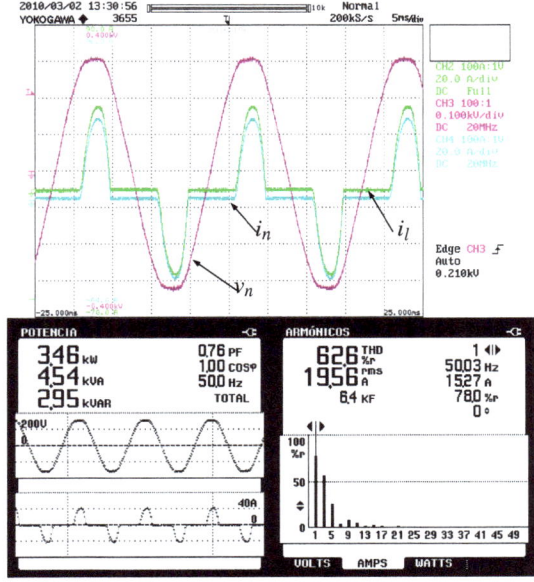

Fig. 7.7 Nonlinear load connected to the AC source working at 50 Hz. Top: v_n, v_1, i_l and i_n. Bottom: PF, $\cos\varphi$ and THD for i_n.

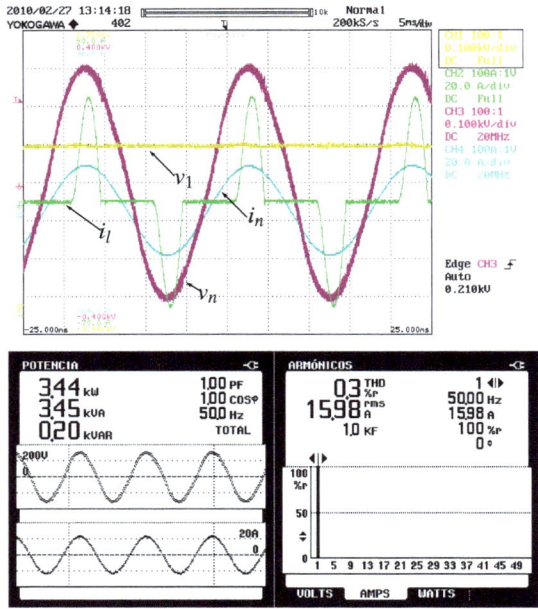

Fig. 7.8 Nonlinear load and the active filter connected to source (50 Hz). Top: v_n, i_n, i_l and v_1; Bottom: PF, $\cos\phi$ and THD for i_n.

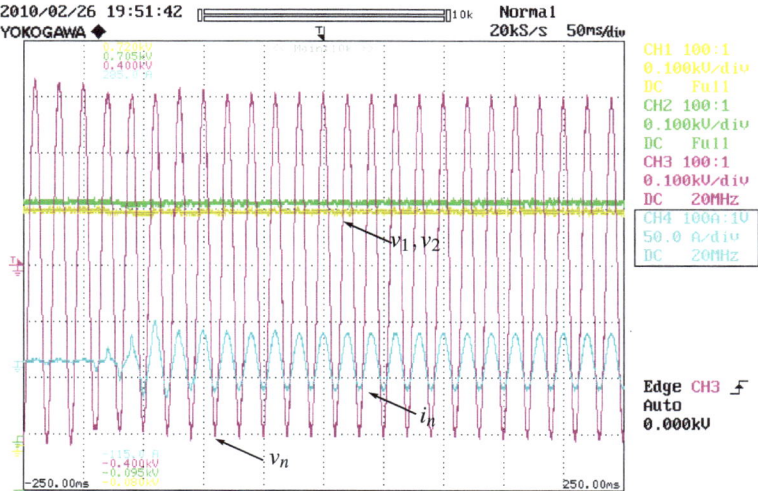

Fig. 7.9 An off-on transition of the nonlinear load with the active filter connected to source (50 Hz). v_n, i_n, v_1 and v_2.

7.2 Odd Harmonic Repetitive Controller 111

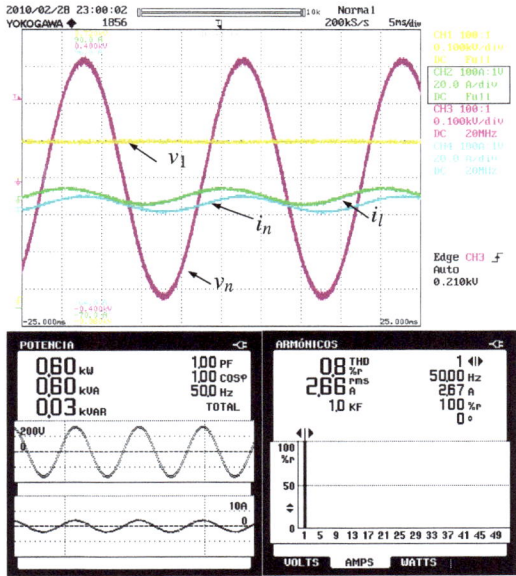

Fig. 7.10 Linear load and the active filter connected to source (50 Hz). Top: v_n, i_n, i_l and v_1; Bottom: PF, $\cos\phi$ and THD for i_n.

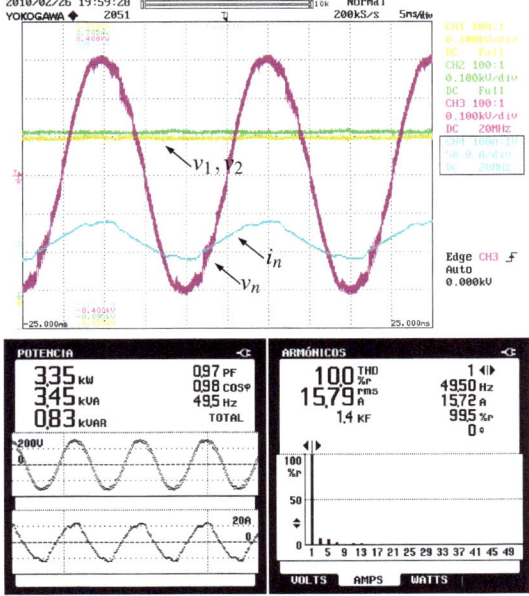

Fig. 7.11 Nonlinear load and the active filter connected to source (49.5 Hz). Top: v_n, i_n, v_1, and v_2; Bottom: PF, $\cos\phi$ and THD for i_n.

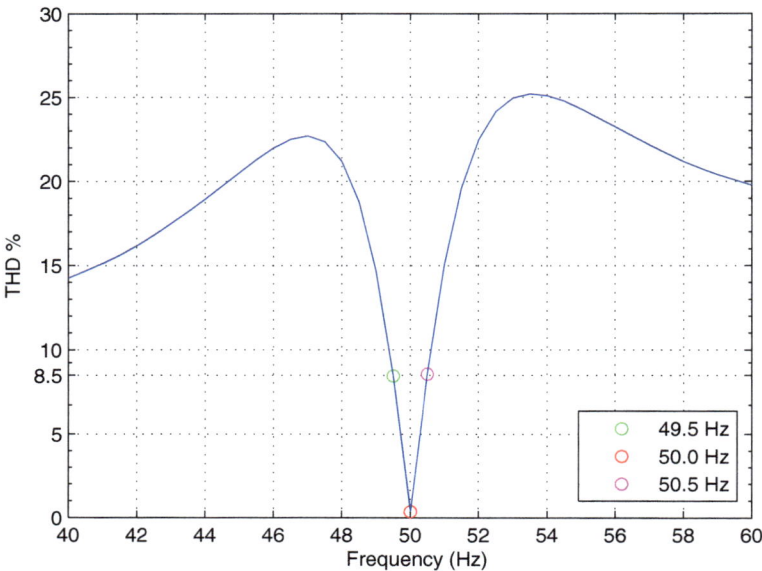

Fig. 7.12 THD degradation due to the frequency variation having a nonlinear load

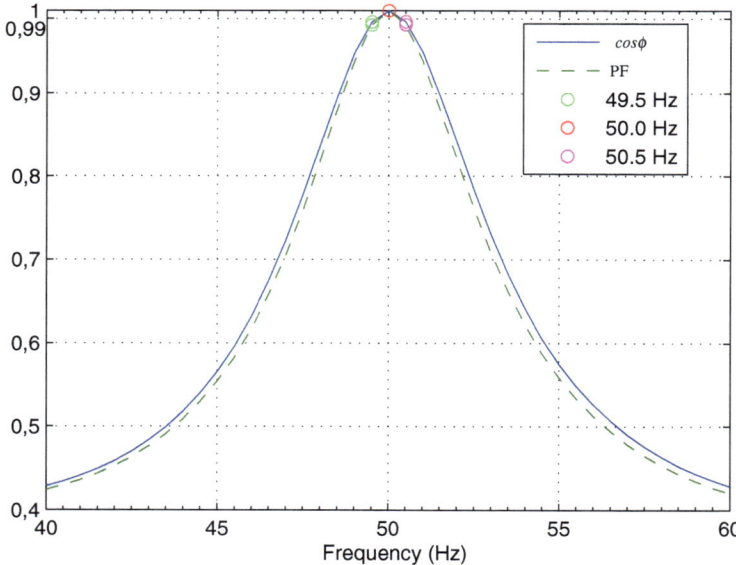

Fig. 7.13 $cos\phi$ and PF degradation due to the frequency variation having a nonlinear load

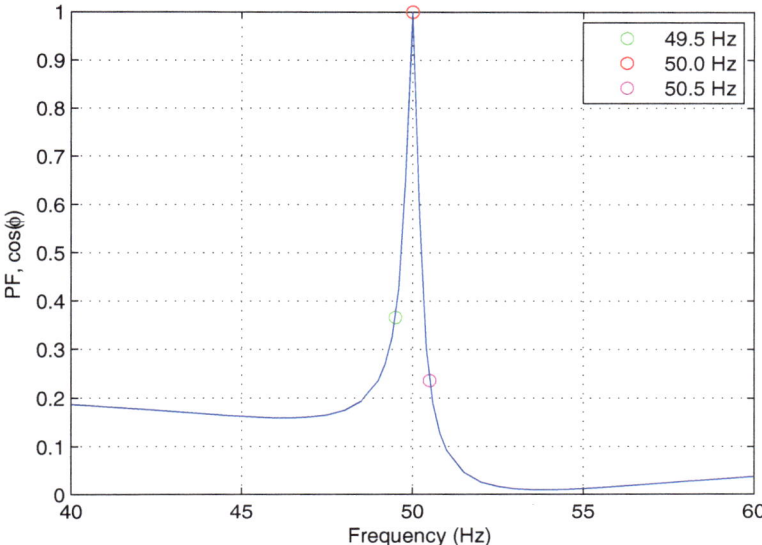

Fig. 7.14 PF variation as a function of the network frequency having a pure capacitive load

7.2.2 *Performance under Network Frequency Variations*

The network frequency is prone to suffer slight variations around its nominal value. Since the standard design of repetitive controllers uses a fixed network frequency, the system performance will be largely degraded at frequency values diverging from the nominal one. For the active filter application, this is reflected in a shape degradation of the source current, which in turn can be seen as the THD, PF and $\cos\phi$ degradation of the system.

In order to show the performance decay under frequency variations, the same controller and settings of the preceding section have been used.

Figure 7.11 shows the response when the frequency is deviated to 49.5 Hz. As depicted, an important shape degradation occurs which can be noticed through the presence of the additional harmonic components. The THD is now 10% and the power factor and $\cos\phi$ are 0.97 and 0.98, respectively.

THD, PF and $\cos\phi$ Degradation

Figures 7.12 and 7.13 show the simulated performance decay through the THD, PF and $\cos\phi$ degradation with respect to the frequency variation. For this computation, the current control loop and the energy control loop have been included in the simulation and the load current $i_l(t)$ shown in Figure 7.8 has been assumed. Notice from Figure 7.12 that, as the network frequency departs from the nominal one the system looses its tracking and rejection capacity allowing the increase of the harmonic components, which is directly reflected in the THD level. It is worth noticing that the simulated THD degradation seems to be lower than the experimental one, which can be explained through the additional degradation of the voltage source in practice, plant non-linearities and unmodeled dynamics. On the other hand, Figure 7.13 shows that although the load has a $cos\phi$ close to 1, the $cos\phi$ and PF exhibit some degradation due to the frequency variation. In case of capacitive or inductive loads, the active filter control system will provide $cos\phi$ and PF correction at nominal frequency. However this performance will be also compromised when the network frequency changes with the subsequent $cos\phi$ and PF degradation. Fig. 7.14 shows the PF behavior using the settings of the previous section with a pure capacitive load. As it can be seen, an important PF degradation occurs when the network frequency deviates from the nominal one.

7.3 Varying Sampling Results

7.3.1 Implementation Issues

The experimental setup consists of the ac power source PACIFIC Smartsource 140-AMX-UPC12 that acts as a variable frequency ac source, a full-bridge diode rectifier acting as a nonlinear load and the previously described single-phase active filter, which is connected in a shunt manner with the rectifier to compensate its nonlinear effects.

The active filter controller has been digitally implemented on a DSP based hardware composed of an ADSP-21161 floating-point DSP processor with an ADSP-21990 fixed-point mixed-signal DSP processor that acts as coprocessor, both from Analog Devices. The ADSP-21161 and the ADSP-21990 communicate with each other using a high-speed synchronous serial channel in Direct Memory Access (DMA) mode. The ADSP-21990 deals with the PWM generation and the A/D conversions with the provided 14 bits eight high-speed A/D channels. The PWM switching frequency is equal to the nominal sampling frequency of 20 kHz.

The network frequency is obtained through some additional hardware that runs in the DSP. Figure 7.15 shows the three main components of the frequency estimation : 1) the zero crossing detector module, which comprises a low-pass filter to remove noise from the voltage measure and a smith trigger comparator to detect the zero crossings; 2) the frequency calculation module, which consists of two counters acting as a period estimation block and, finally, 3) a low-pass digital filter in charge of reducing noise in the digital estimation of the period. With this information the sampling frequency is updated to maintain the ratio $N = 400$.

It is well known that in a real setup T_s cannot be fixed with infinite precision but within the limitations imposed by the timer quantification. Notice that even small quantification errors entail important gain reductions.

In spite of the gain reduction, the sampling time adaptation scheme allows to maintain the gain of (7.12) above 52 dB, which is good enough for most applications.

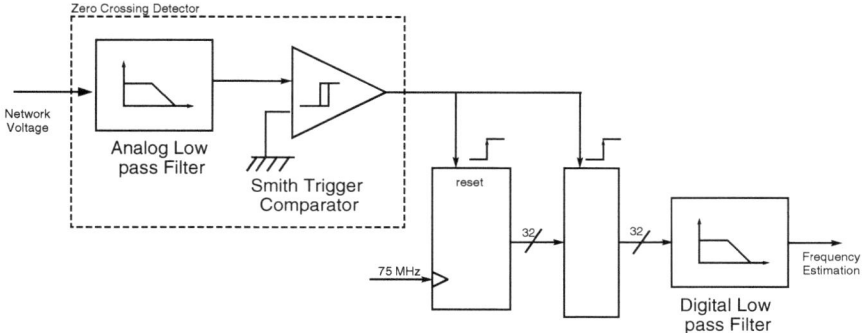

Fig. 7.15 Network frequency computation

7.3.2 Robust Control Theory Approach

According to Section 3.4 and Remark 3.3, the previous settings yield $\|G_{\bar{T}}(z)\|_\infty = 2.6530 \times 10^4$. In order to define $\gamma_{\bar{T}}$ (see (3.16)) we select, $\varepsilon = 0.0001$. Moreover, the continuous-time plant matrix is

$$A = \begin{pmatrix} -28651.9058 & -4276.5664 \\ 4096 & 0 \end{pmatrix}.$$

Table 7.1 collects the stability intervals obtained with the optimized Γ defined in (3.18) and also for $\Gamma = \mathbb{I}$ (i.e. the proposal in [10, 20]) using three different norm bounds for the matrix exponential Δ: a numerical calculation, the log norm of a matrix with respect to the 2-norm [10] and a Schur decomposition-derived bound [20] (see Appendix B for more details about the calculation of the stability intervals). Notice that the numerical bound and the log norm provide similar results, while the latter appears to be sharper than the Schur decomposition bound. The conservatism reduction achieved with the assignment of Γ proposed in this article is between 300% and 336%, depending on the norm bound calculation method. It is worth mentioning that nominal line frequency is usually 50 Hz or 60 Hz depending on the geographical area, and may vary by not more than 10 % of its value. Therefore, one can expect line frequency variations in the interval $[45, 66]$ Hz.

Table 7.1 Shunt active filter: stability intervals in frequency units (Hz)

Bounds	$\Gamma = \mathrm{col}([0,0]^\top, [0,\mathbb{I}]^\top)$	$\Gamma = \mathbb{I}$
Numerical	[28.4944, 102.3305]	[42.4676, 59.3855]
Log norm	[28.5096, 102.2870]	[42.4680, 59.3850]
Schur dec.	[29.9672, 95.5837]	[42.6741, 59.0878]

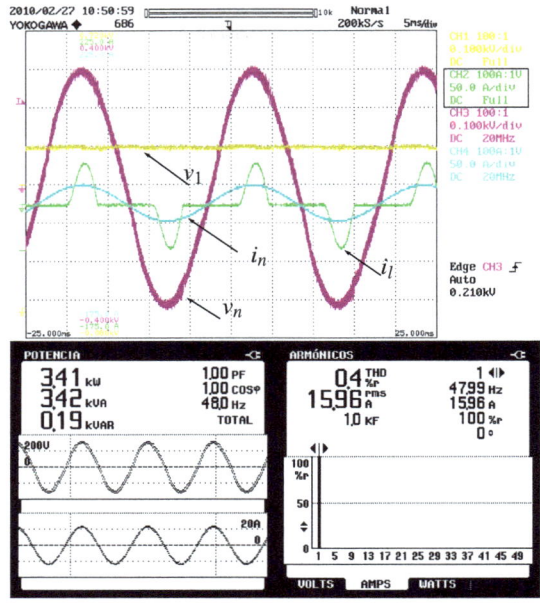

Fig. 7.16 Nonlinear load with active filter connected to the ac source working at 48 Hz with adaptive scheme. Top: v_n, v_1, i_l and i_n. Bottom: PF, $\cos\varphi$ and THD for i_n.

7.3.3 Experimental Results

Figure 7.16 shows that when the frequency of the voltage source is fixed to 48 Hz and the active filter is connected in parallel with the rectifier, the shape of the current at the source port is nearly sinusoidal with a THD of 0.4%. Moreover, the PF and $\cos\phi$ at the port are unitary.

Figure 7.17 portrays the system response when the network frequency is 52 Hz and the sampling time is adapted as described in this work. As it can be seen, the system behavior is similar to the one obtained when working at nominal frequency (shown in Figure 7.8).

In the next experiment the network frequency is changed from 53 Hz to 48 Hz in a ramp manner (40 cycles), the nonlinear load being also in use. Figure 7.18 depicts the responses of v_n and i_n. Notice that, after a transient, the system reaches the steady-state maintaining the performance.

For the next two experiments the linear load is used and the network frequency of the system is changed to 51 Hz and 49 Hz, while the sampling time is adapted as previously described. Figures 7.19 and 7.20 show that the system performance is kept similar to the one obtained for the nominal frequency providing unitary PF and $\cos\phi$.

7.3 Varying Sampling Results 117

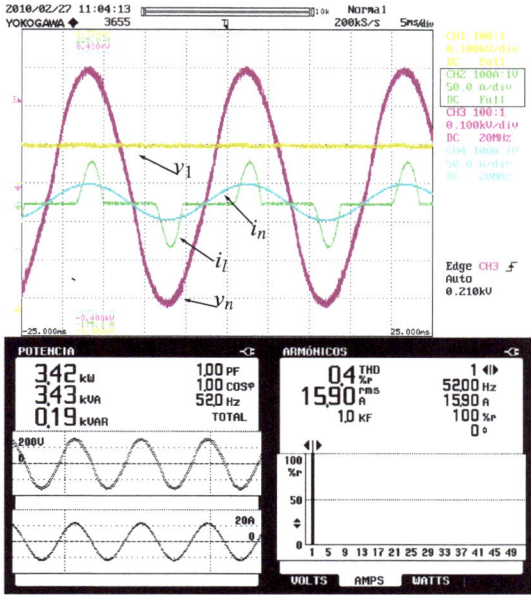

Fig. 7.17 Nonlinear load with active filter connected to the ac source working at 52 Hz with adaptive scheme. Top: v_n, v_1, i_l and i_n. Bottom: PF, cos φ and THD for i_n.

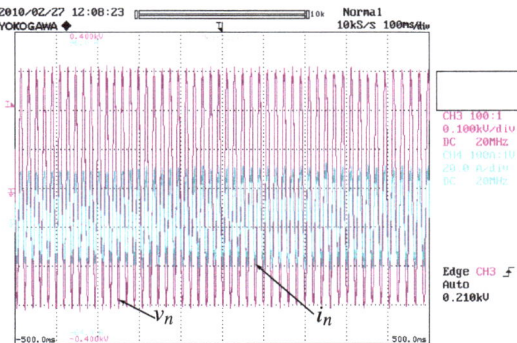

Fig. 7.18 Nonlinear load with active filter connected to the ac source with adaptive scheme. v_n and i_n when the network frequency changes from 53 Hz to 48 Hz in 49 cycles.

118   7 Shunt Active Power Filter

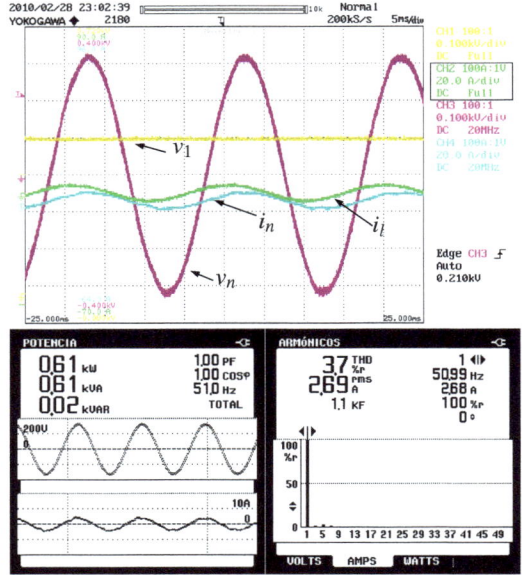

Fig. 7.19 Linear load and the active filter connected to source (51 Hz). Top: v_n, i_n, i_l and v_1; Bottom: PF, $\cos\phi$ and THD for i_n.

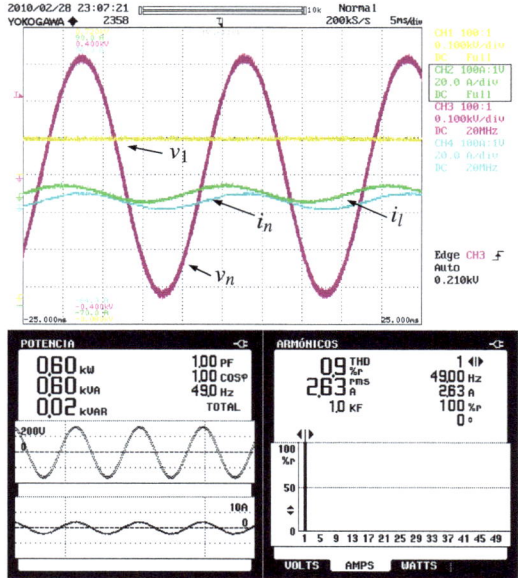

Fig. 7.20 Linear load and the active filter connected to source (49 Hz). Top: v_n, i_n, i_l and v_1; Bottom: PF, $\cos\phi$ and THD for i_n.

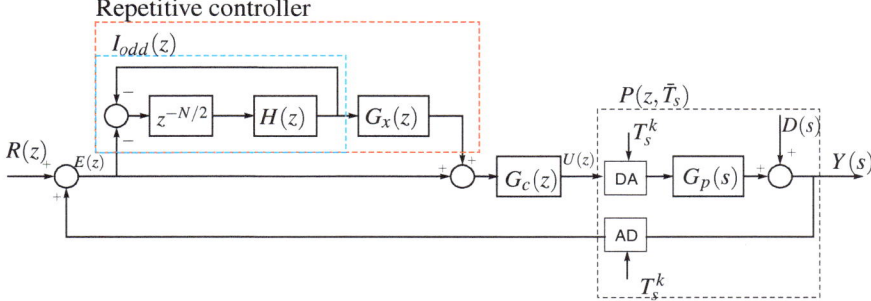

Fig. 7.21 Discrete-time block-diagram of the basic varying sampling repetitive control structure

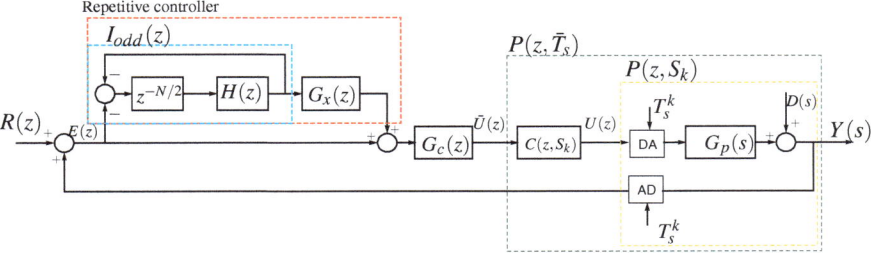

Fig. 7.22 Discrete-time block-diagram of the closed-loop system with the adaptation-compensation controller structure

7.4 Adaptive Pre-compensation

7.4.1 Controller Design

Aiming at annihilating the effect of the time-varying sampling and forcing an output behavior corresponding to that of the nominal sampling period $\bar{T}$, a pre-compensator is introduced between the nominal controller $G_c(z)$ and the plant (see Section 4.3). Namely, let the repetitive controller be designed and implemented to provide closed-loop stability for an a priori selected nominal sampling period $\bar{T}$, using the LTI model

$$G_p(z,\bar{T}) \triangleq \frac{Num(z,\bar{T})}{Den(z,\bar{T})} = \mathscr{L}\{G_p(s)\}$$

(see Figure 7.21), with

$$G_p(s) = \frac{i_f(s)}{\alpha(s)} = -\frac{\frac{1}{r_l}}{\left(\frac{L}{r_L}s+1\right)(\tau s+1)},$$

120   7 Shunt Active Power Filter

τ being the time constant of the antialiasing filter. When working at varying sampling period T_s, this model is a LTV system

$$G_p(z, T_s) = \frac{Num(z, T_s)}{Den(z, T_s)},$$

with $T_s = \{T_k, T_{k-1}, T_{k-2}\}$ (recall that the order of $G_p(s)$ is 2). In order to annihilate the effect of the sampling rate change, the pre-compensator

$$C(z, T_s) = G_p(z, \bar{T}) G_p^{-1}(z, T_s) = \frac{Num(z, \bar{T})}{Den(z, \bar{T})} \frac{Den(z, T_s)}{Num(z, T_s)} \quad (7.14)$$

is connected in series with the LTV plant $G_p(z, T_s)$. Thence, the overall behaviour is that of the nominal LTI system: $C(z, T_s) G_p(z, T_s) = G_p(z, \bar{T})$. Figure 7.22 inner loop control scheme.

As the pre-compensator-plant subsystem is kept invariant and equal to the nominal plant, closed-loop system stability is preserved. Additionally, the inner loop transfer function is preserved invariant and, as a consequence, outer-loop stability is also preserved. The pre-compensator defined in (7.14) can be equivalently implemented in input-output or state-space approaches, but the state-space formulation leads to a more efficient code. Finally, although this approach guarantees closed-loop stability, it is necessary to check the internal stability of the compensator-plant subsystem $P(z)$ (including possible forbidden cancellations). LMI gridding techniques (see Section 3.3) allow to prove that, for the active filter used in this paper, the pre-compensation scheme is internally stable for a wide range of values of T_s, in particular those covering the most relevant frequency interval from the practical point of view ([45, 66] Hz).

7.4.2 Controller Calculation

The active filter system together with the anti-aliasing filter can be seen as a second order continuous-time system with the following transfer function

$$G(s) = \frac{k_1}{\tau_1 s + 1} \frac{k_2}{\tau_2 s + 1}, \quad (7.15)$$

where $k_1 = -1/r_L$, $\tau_1 = L/r_L$, $k_2 = 1$, and τ_2 being the time constant of the anti-aliasing filter. A state-space Jordan form is:

$$\dot{x}(t) = \begin{bmatrix} -\frac{1}{\tau_1} & 1 \\ 0 & -\frac{1}{\tau_2} \end{bmatrix} x(t) + \begin{bmatrix} 0 \\ 1 \end{bmatrix} u(t)$$

$$y(t) = \begin{bmatrix} \frac{k_1 k_2}{\tau_1 \tau_2} & 0 \end{bmatrix} x(t). \quad (7.16)$$

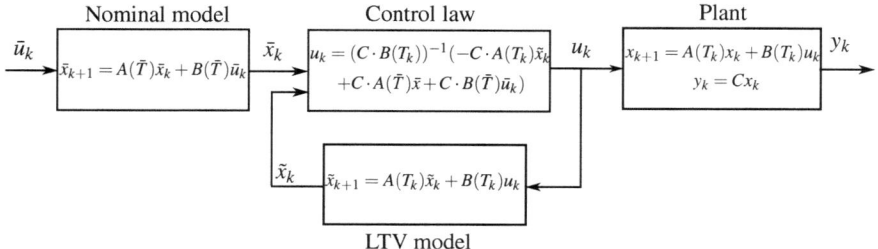

Fig. 7.23 Pre-filtering scheme

The discretization of the continuous model (7.16) for a given T_k is

$$x_{k+1} = A(T_k)x_k + B(T_k)u_k \tag{7.17}$$
$$y_k = Cx_k, \tag{7.18}$$

with

$$A(T_k) = \begin{bmatrix} e^{-\frac{T_k}{\tau_1}} & \frac{\left(-e^{-\frac{T_k}{\tau_2}} + e^{-\frac{T_k}{\tau_1}}\right)\tau_2\tau_1}{(\tau_1 - \tau_2)} \\ 0 & e^{-\frac{T_k}{\tau_2}} \end{bmatrix},$$

$$B(T_k) = \begin{bmatrix} \frac{\tau_2\tau_1\left(\tau_1 e^{-\frac{T_k}{\tau_1}} - \tau_2 e^{-\frac{T_k}{\tau_2}} - \tau_1 + \tau_2\right)}{(\tau_1 - \tau_2)} \\ -\tau_2\left(e^{-\frac{T_k}{\tau_2}} - 1\right) \end{bmatrix},$$

$$C = \begin{bmatrix} \frac{k_1 k_2}{\tau_1 \tau_2} & 0 \end{bmatrix}.$$

The model of the plant at a given nominal sampling period $\bar{T}$ is

$$\bar{x}_{k+1} = A(\bar{T})\bar{x}_k + B(\bar{T})\bar{u}_k \tag{7.19}$$
$$\bar{y}_k = C\bar{x}_k. \tag{7.20}$$

Control Law

The goal is to find u_k such that $y_k = \bar{y}_k$ or, equivalently, $Cx_k = C\bar{x}_k$. With this, and the discrete LTV plant (7.17) the expression for $y_{k+1} = \bar{y}_{k+1}$ is obtained:

$$C \cdot A(\bar{T})\bar{x}_k + C \cdot B(\bar{T})\bar{u}_k = C \cdot A(T_k)x_k + C \cdot B(T_k)u_k. \tag{7.21}$$

Thus, the control law becomes[7]:

$$u_k = (C \cdot B(T_k))^{-1}(-C \cdot A(T_k)x_k + C \cdot A(\bar{T})\bar{x} + C \cdot B(\bar{T})\bar{u}_k). \tag{7.22}$$

---

[7] In the case under study $(C \cdot B(T_k))^{-1}$ always exists and it is well-defined.

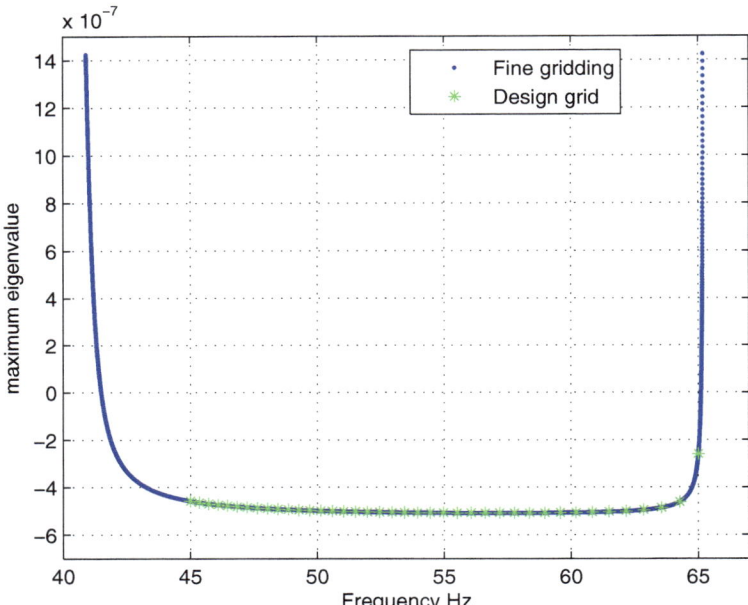

Fig. 7.24 Internal stability evaluation. Eigenvalues evolution of $L_{T_k}(P_G)$ using 40 points in the design grid and 50000 points to check stability.

Since the state of the plant is not available we can use the following LTV model to obtain an estimation $\tilde{x}_k$ of the state x_k:

$$\tilde{x}_{k+1} = A(T_k)\tilde{x}_k + B(T_k)u_k \qquad (7.23)$$
$$y_k = C\tilde{x}_k. \qquad (7.24)$$

Then the control law is:

$$u_k = (C \cdot B(T_k))^{-1}(-C \cdot A(T_k)\tilde{x}_k + C \cdot A(\bar{T})\bar{x} + C \cdot B(\bar{T})\bar{u}_k). \qquad (7.25)$$

Figure 7.23 shows the block diagram of this control law. The dynamics of the system composed of the pre-compensator and the plant is:

$$\begin{bmatrix} \bar{x} \\ \tilde{x} \\ x \end{bmatrix}_{k+1} = \begin{bmatrix} A(\bar{T}) & 0 & 0 \\ B(T_k)(C \cdot B(T_k))^{-1}C \cdot A(\bar{T}) & -B(T_k)(C \cdot B(T_k))^{-1}C \cdot A(T_k) + A(T_k) & 0 \\ B(T_k)(C \cdot B(T_k))^{-1}C \cdot A(\bar{T}) & -B(T_k)(C \cdot B(T_k))^{-1}C \cdot A(T_k) & A(T_k) \end{bmatrix} \begin{bmatrix} \bar{x} \\ \tilde{x} \\ x \end{bmatrix}_k$$
$$+ \begin{bmatrix} B(\bar{T}_k) \\ B(T_k)(C \cdot B(T_k))^{-1}C \cdot B(\bar{T}) \\ B(T_k)(C \cdot B(T_k))^{-1}C \cdot B(\bar{T}) \end{bmatrix} \bar{u}_k$$
$$y_k = \begin{bmatrix} 0 & 0 & C \end{bmatrix} \begin{bmatrix} \bar{x} \\ \tilde{x} \\ x \end{bmatrix}_k.$$

$$(7.26)$$

Internal Stability

We are interested in analyzing the stability of (7.26) for all sampling periods $T_k \in \mathscr{T}$. The LMI gridding approach introduced in [2, 17] allows a simplified stability analysis that may be performed as follows. Let $\{\tau_0, \ldots, \tau_r\}$, be a sorted set of sampling periods candidate suitably distributed in $\mathscr{T}$. Then, one may solve the following finite set of LMIs:

$$L_{\tau_i}(P) \leq -\alpha \mathbb{I}, \; i = 0, \ldots, r, \; \text{s.t.} \; P = P^\top > 0, \tag{7.27}$$

for a fixed $\alpha \in \mathbb{R}^+$. In case that the problem is feasible and a solution, $P = P_G$, is encountered, the negative-definite character of $L_{T_k}(P_G)$ is to be checked for intermediate values of T_k in each open subinterval (τ_i, τ_{i+1}). If this fails to be accomplished, (7.27) has to be solved again for a finer grid of $\mathscr{T}$. Otherwise, the procedure should be relaunched for a new interval $\mathscr{T}' \subset \mathscr{T}$.

In practice, it would be necessary to assure stability in the frequency interval $[45, 65]$ Hz. A line frequency variation of $\pm 10\%$ of the nominal value is assumed, which encompasses the requirements in international standards and many practical scenarios [13, 21, 22]. Indeed, frequency fluctuations up to $\pm 6\%$ of the nominal value are considered in the literature for different scenarios such as, for example, wind farms [3, 21], photo-voltaic generators [1] and microgrids [8], the usual being $\pm 2\%$ and below. Therefore, recalling the relation $T_n = N \cdot T_s$, one finds out that $\mathscr{T} = \left[(55N)^{-1}, (45N)^{-1} \right]$. However, it is worth pointing out that wider frequency variations may occur in specific situations, such as island grids [21]. In this case the internal stability should be verified according to the expected frequency variation interval.

Thus, considering the here expected variation interval, 40 uniformly distributed points are selected in $\mathscr{T} = \left[5.55 \cdot 10^{-5}, 3.8461 \cdot 10^{-5} \right]$ s. These points are used to construct the set of LMIs (7.27), and a feasible solution $P = P_G$ with $\alpha = 100$ is obtained. Figure 7.24 depicts the maximum modulus eigenvalue of $L_{T_k}(P_G)$, detailing with a star the 40 points leading to the LMI formulation. The maximum modulus eigenvalue of $L_{T_k}(P_G)$ corresponding to a finer grid consisting of 50000 uniformly distributed point are also drawn in Figure 7.24. These points are used to check the sign of $L_{T_k}(P_G)$ in the intervals between the points defining the LMI set. It can be seen that $L_{T_k}(P_G) < 0$ for every point in this finer grid of the interval $\mathscr{T}$; hence, stability is dynamically preserved therein.

7.4.3 Experimental Results

With this information, the sampling period T_s is updated to maintain the ratio $N = 400$. Figure 7.25 and 7.26 show the steady-state behaviour in case of the nonlinear load at 48 Hz and 53 Hz, respectively: the shape of the source current is nearly sinusoidal with a THD of 0.4 and $cos\phi$ at the port are unitary. Figure 7.27 shows the system behaviour when the source frequency changes from 48 Hz to 53 Hz in a 40 cycles ramp manner. Notice that the source current is preserving the required sinusoidal shape and adapting to the variable source voltage frequency.

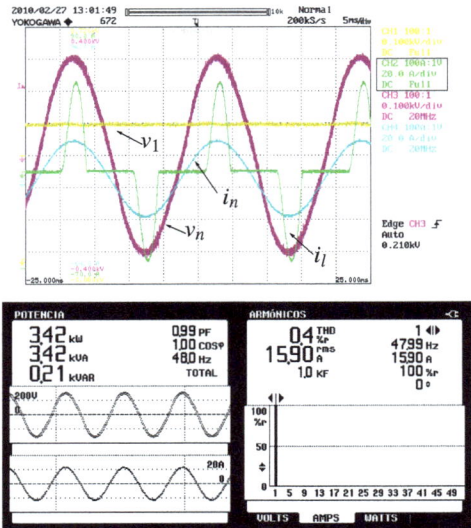

Fig. 7.25 Nonlinear load with active filter connected to the ac source working at 48 Hz with pre-compensation scheme. Top: v_n, v_1, i_l and i_n. Bottom: PF, $\cos\varphi$ and THD for i_n.

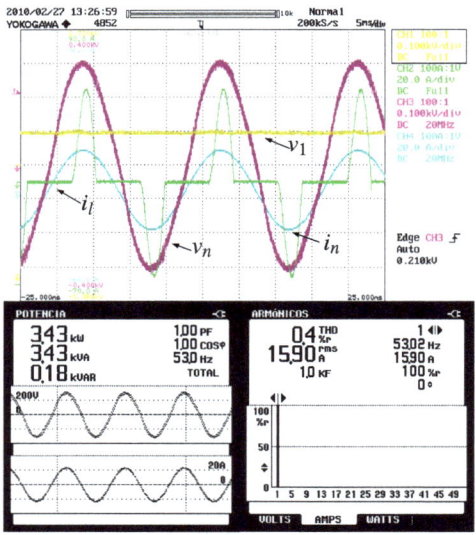

Fig. 7.26 Nonlinear load with active filter connected to the ac source working at 53 Hz with pre-compensation scheme. Top: v_n, v_1, i_l and i_n. Bottom: PF, $\cos\varphi$ and THD for i_n.

7.4 Adaptive Pre-compensation 125

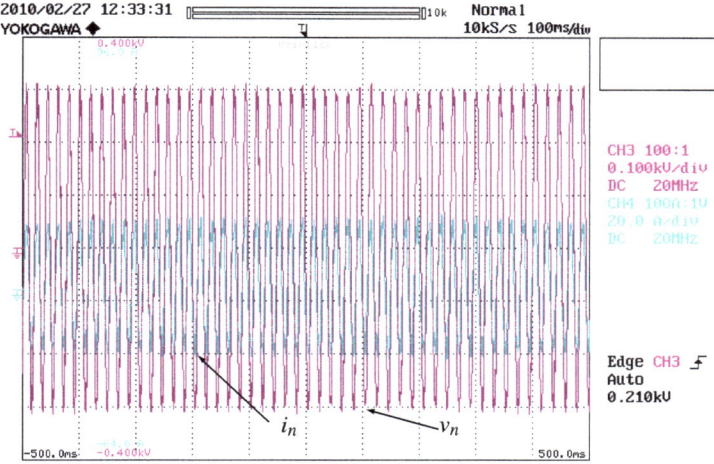

Fig. 7.27 Nonlinear load with active filter with pre-compensation scheme when the network frequency changes from 48Hz to 53Hz: v_n, i_n

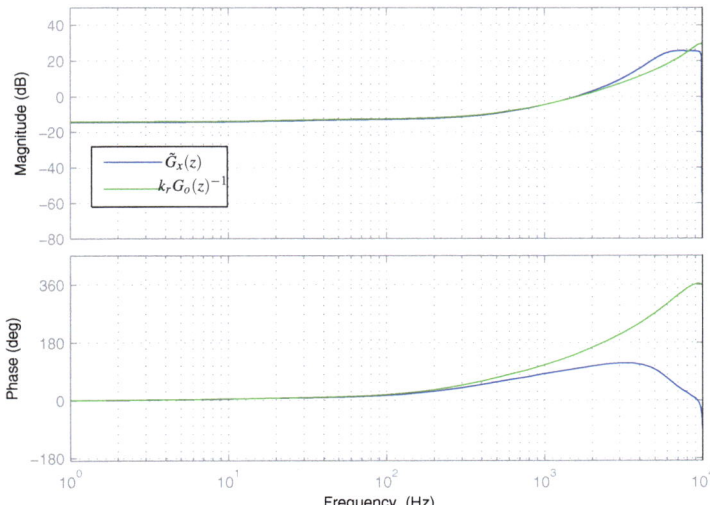

Fig. 7.28 Frequency response comparison between $\tilde{G}_x(z)$ and of the original controller $G_x(z) = k_r G_o^{-1}(z)$, with $k_r = 0.197$

7.5 Robust Design

A μ-synthesis approach can take advantage from the problem structure. Using the latter, the following controller is obtained for $T_s \in \mathcal{T} = \left[(55N)^{-1}, (45N)^{-1}\right]$, with $N = 400$:

$$\tilde{G}_x(z) = \frac{3.182z^7 - 6.432z^6 + 0.3002z^5 + 6.54z^4 - 3.791z^3 - 0.1361z^2 + 0.3089z + 0.02804}{z^7 - 0.2715z^6 - 1.165z^5 - 0.1143z^4 + 0.04983z^3 + 0.3138z^2 + 0.1596z + 0.02804}. \tag{7.28}$$

Remark 7.1. Figure 7.28 shows that, for $T_s = \bar{T}$, the frequency responses of $\tilde{G}_x(z)$ and of the original controller, i.e. $G_x(z) = k_r G_o^{-1}(z)$, with $k_r = 0.197$ instead of the original $k_r = 0.7$, are very similar in the low and medium range[8]. Hence, aiming at maintaining the controller order as low as possible, in this case the practical implementation maintains the original structure for $G_x(z)$, with the adjusted parameter value $k_r = 0.197$.

It is worth to note that the μ-synthesis design seeks a causal filter $\tilde{G}_x(z)$ for a given frequency variation interval, while the standard design, for $G_o(z)$ being a minimum phase system, uses $G_x = k_r/G_o(z)$ which is a non causal transfer function.

Finally, it is worth recallling that, according to the IMP, steady-state performance is guaranteed when T_n (and, consequently, T_s) remains constant for sufficiently large time intervals.

Figures 7.29 and 7.30 show the open-loop and sensitivity function frequency response of the odd-harmonic RC designed in Section 7.2 and the robust design of this section using $\tilde{G}_x(z)$. It can be seen that the tracking/rejection action is lower when using $\tilde{G}_x(z)$ as it represents a lower gain k_r. Additionally, $\tilde{G}_x(z)$ produces a sensitivity function frequency response with better inter-harmonic behavior.

7.5.1 Experimental Results

Figures 7.31 and 7.32 show the system response when the network frequency is 49 Hz and 51 Hz, respectively, and the sampling period of the active filter repetitive controller is adapted as described in this Section. Notice that the performances are similar to the ones reported in Subsection 7.2.1 when both the network frequency and the controller sampling period work at nominal values (see Figure 7.8).

The last experiment consist of a step change in frequency from 50 Hz to 51 Hz. Figure 7.33 shows that after a small transient the system performance is reached again.

---

[8] In this application, it is expected to deal with currents with relevant harmonic components until the 13th or 15th harmonics, which corresponds to 650 Hz and 750 Hz. Therefore, harmonic compensation is intended to take place within this region, which is considered as the medium range in this remark

7.5 Robust Design

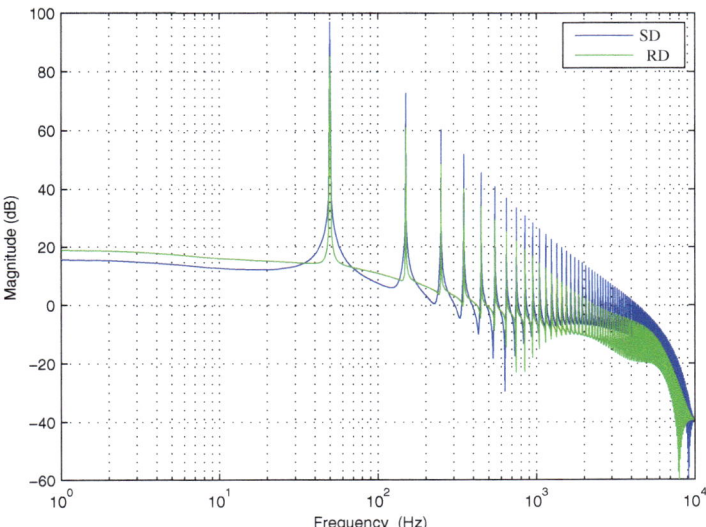

Fig. 7.29 Open-loop magnitude response of function $G_l(z)$. Standard design (SD) and robust design (RD) comparison.

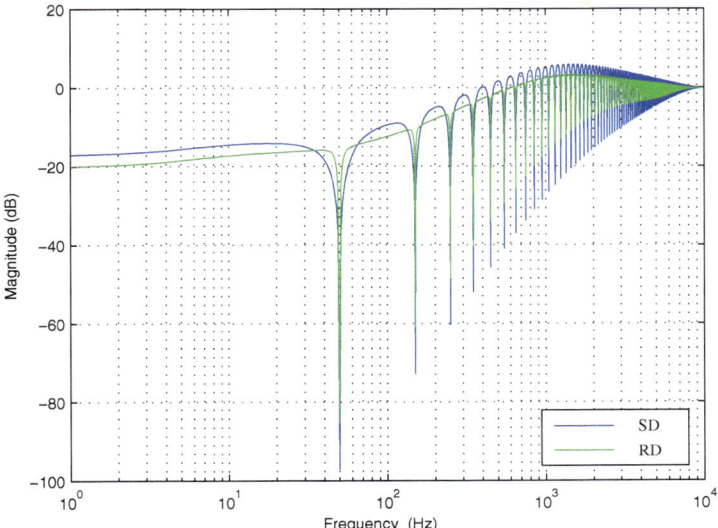

Fig. 7.30 Sensitivity function magnitude response. Standard design (SD) and robust design (RD) comparison.

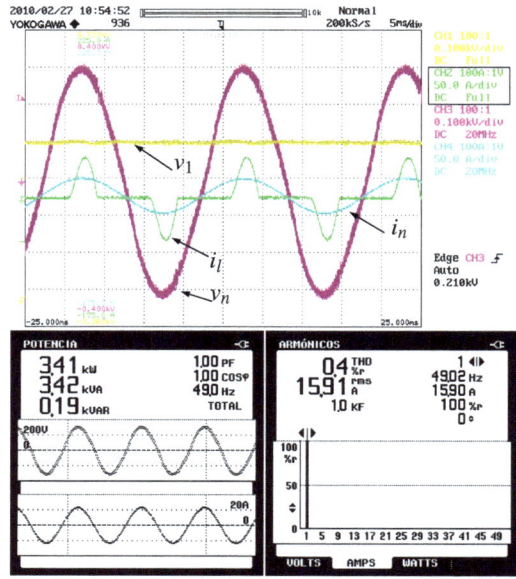

Fig. 7.31 Nonlinear load with active filter connected to the ac source working at 49 Hz using the robust design. Top: v_n, v_1, i_l and i_n. Bottom: PF, $\cos\varphi$ and THD for i_n.

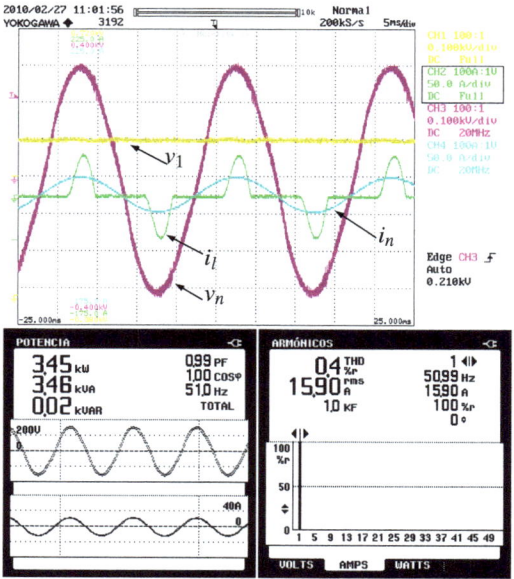

Fig. 7.32 Nonlinear load with active filter connected to the ac source working at 51 Hz using the robust design. Top: v_n, v_1, i_l and i_n. Bottom: PF, $\cos\varphi$ and THD for i_n.

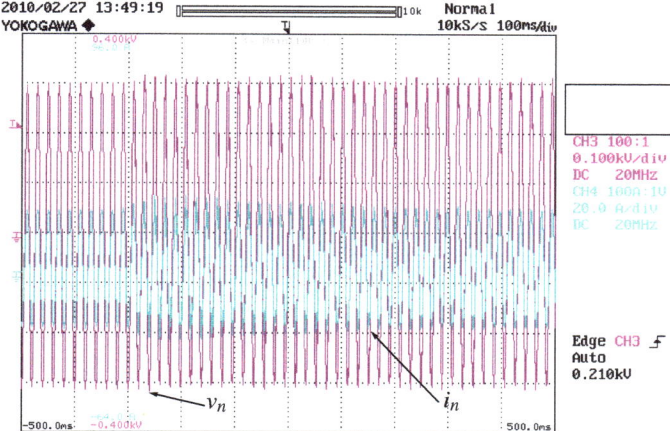

Fig. 7.33 Nonlinear load with active filter connected to the ac source using the robust design. v_n and i_n with a step change in the network frequency from 50 Hz to 51 Hz.

7.6 HORC

7.6.1 Experimental Setup

The settings are the same used in Section 7.2.1 but setting $k_r = 1$, changing the standard IM by the second order odd-harmonic IM described in Section 5.3.5:

$$I_{hodd}(z) = -\frac{\left(2z^{-\frac{N}{2}} + z^{-N}\right)H(z)}{1 + \left(2z^{-\frac{N}{2}} + z^{-N}\right)H(z)}$$

and defining

$$H(z) = 0.06241z^{-3} + 0.1293z^{-2} + 0.1963z^{-1} + 0.2239 + 0.1963z + 0.1293z^2 + 0.06241z^3,$$

Remark 7.2. In the active filter application, it is desirable to separate the dynamics of the two existing loops: the current loop and the energy shaping loop, where the current loop must be the fastest one. In this way the second-order odd-harmonic internal model, which is slower than the standard one, should be designed as fast as possible, which is done using $k_r = 1$.

With this controller, the obtained open-loop and sensitivity function magnitude response are shown in Figure 7.34 and 7.35, respectively. Also a comparison with the odd-harmonic repetitive control designed in Section 7.2 is shown in these Figures. Both show the robustness improvement in the face of frequency variations attained by the odd-harmonic HORC. In Figure 7.35, it is shown that the sensitivity function of the odd-harmonic HORC has a higher maximum gain in the high frequency

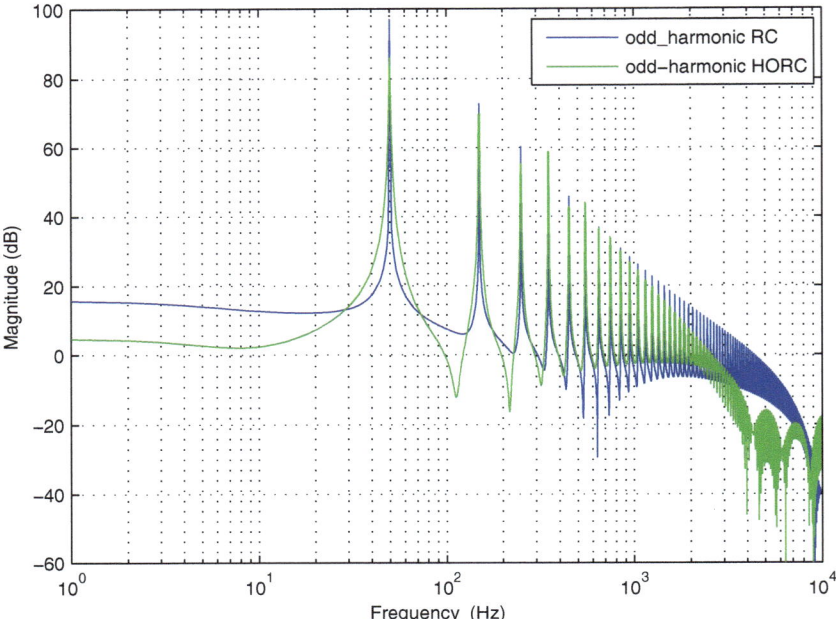

Fig. 7.34 Open-loop magnitude response of function $G_l(z)$. Odd-harmonic RC and HORC design comparison.

range than the odd-harmonic RC, which can be considered as the cost of enhancing robustness. However, the filter $H(z)$ has been designed to exhibit a lower cut-off frequency, so as to avoid problems due to the high gain in the highest frequency interval.

7.6.2 Experimental Results

In the first experiment a rectifier is connected to the AC source, v_s, which is set to 50 Hz. The rectifier current, i.e. the load current i_l, has a THD of 62.6% and an RMS value of 19.56A. As shown in Figure 7.36, with the active filter connected in parallel with the rectifier, the current at the source port, i_n, shows a nearly sinusoidal shape, with a THD of 0.6%, while PF and $\cos\phi$ at the port are unitary. The figure also shows that the mean value of v_1 is maintained almost constant[9]. In the next experiment the same nonlinear load is plugged to the ac source, v_s, now set to 49.5 Hz. As Figure 7.37 shows, the PF and the $\cos\phi$ remain close to unitary values and the THD for i_n is 1.7%. Although there is a small degradation, it is considerably

---

[9] v_2 is not shown due the limited number of channels in the instrumentation.

7.6 HORC 131

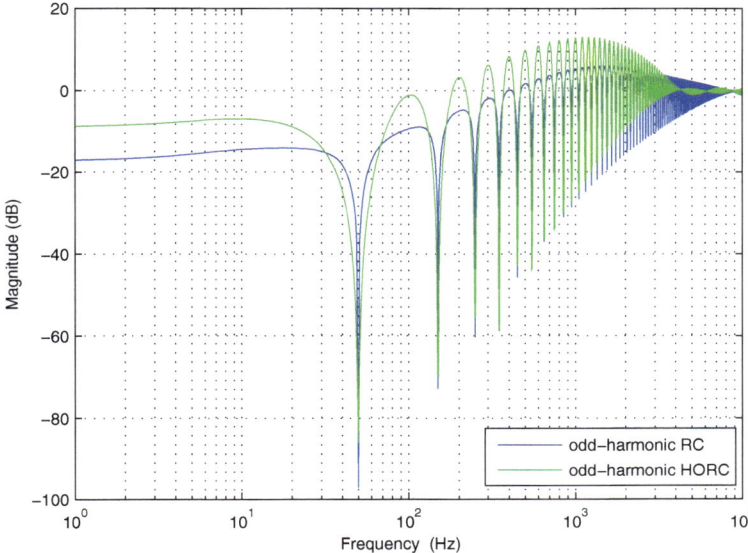

Fig. 7.35 Sensitivity function magnitude response. Odd-harmonic RC and HORC design comparison.

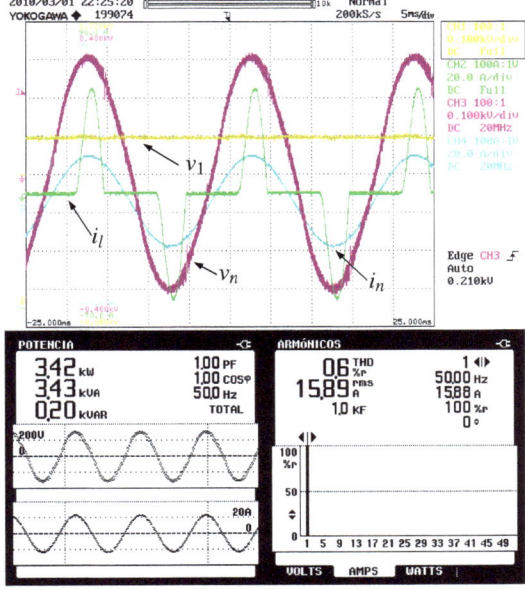

Fig. 7.36 Nonlinear load and the active filter connected to source (50 Hz) using the odd-harmonic HORC. (top) v_n, i_n, i_l and v_1 vs time; (bottom) PF, $\cos\phi$ and THD for i_n.

132   7 Shunt Active Power Filter

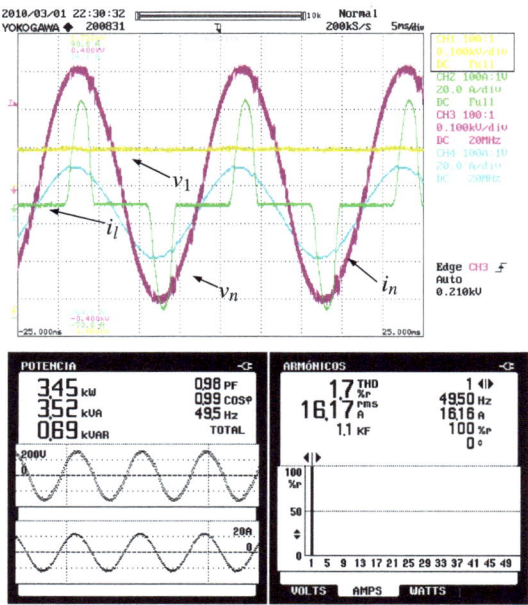

Fig. 7.37 Nonlinear load and the active filter connected to source (49.5 Hz) using the odd-harmonic HORC. (top) v_n, i_n, i_l and v_1 vs time; (bottom) PF, $\cos\phi$ and THD for i_n.

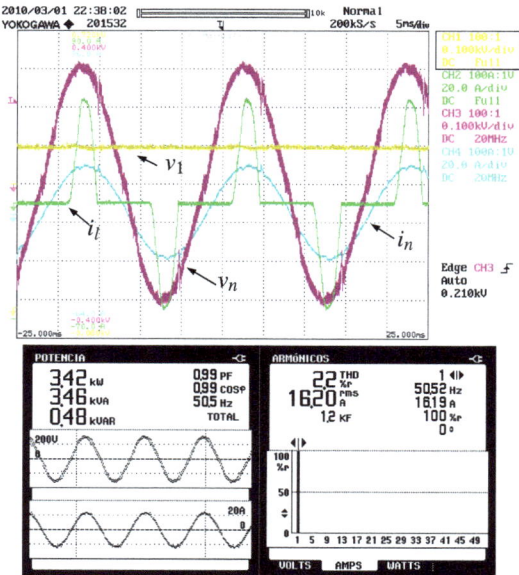

Fig. 7.38 Nonlinear load and the active filter connected to source (50.5 Hz) using the odd-harmonic HORC. (top) v_n, i_n, i_l and v_1 vs time; (bottom) PF, $\cos\phi$ and THD for i_n.

7.6 HORC   133

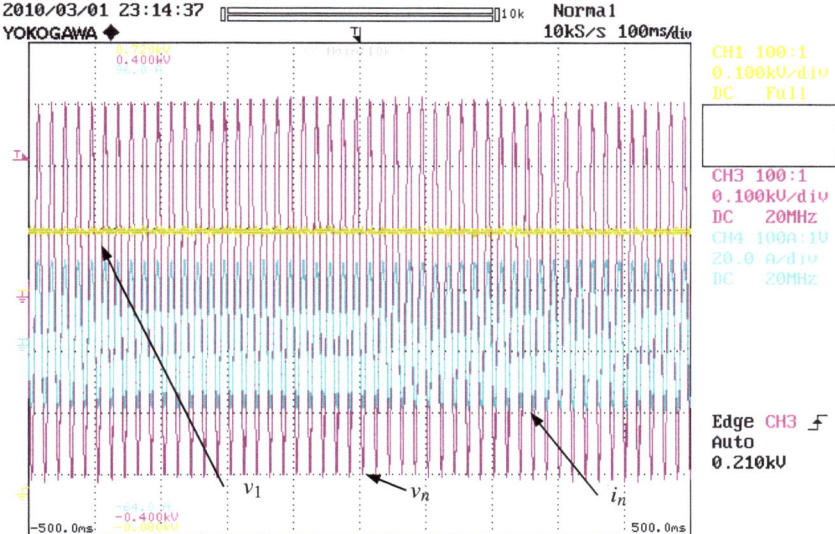

Fig. 7.39 Nonlinear load and the active filter connected to source using the odd-harmonic HORC. Frequency transition from 49.5 Hz to 50.5 Hz. v_n, i_n, and v_1 vs time.

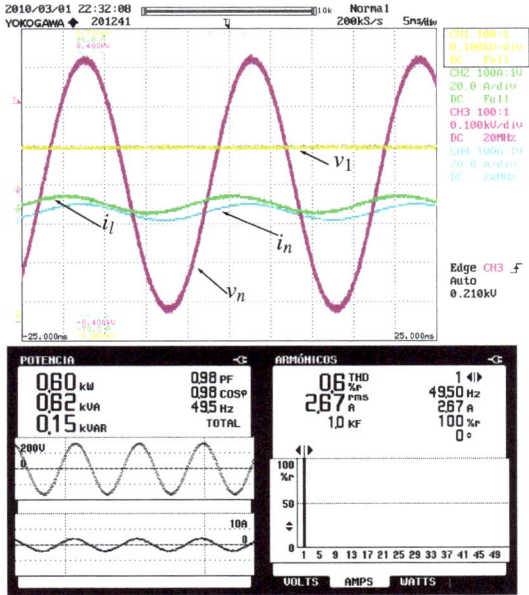

Fig. 7.40 Linear load and the active filter connected to source using the odd-harmonic HORC (49.5 Hz). Top: v_n, i_n, i_l and v_1; Bottom: PF, $\cos\phi$ and THD for i_n.

134   7 Shunt Active Power Filter

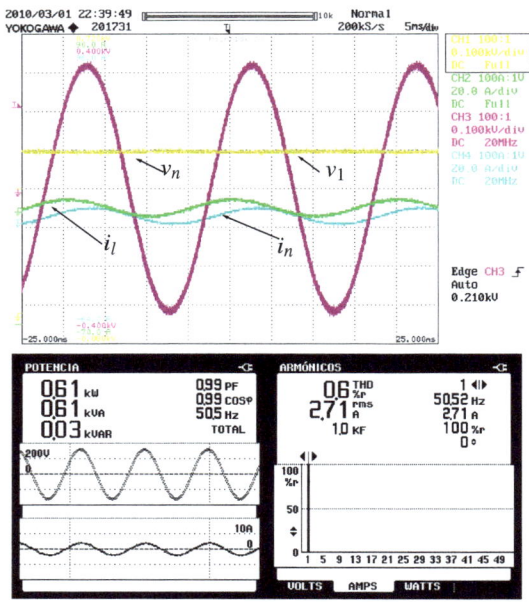

Fig. 7.41 Linear load and the active filter connected to source using the odd-harmonic HORC (50.5 Hz). Top: v_n, i_n, i_l and v_1; Bottom: PF, $\cos\phi$ and THD for i_n.

lower than the one presented by the standard RC (see Figure 7.11) which shows the successful robustness improvement. Figure 7.38 shows the results with the source configured to 50.5 Hz. Notice that, also in this case the degradation is kept small with the PF and the $\cos\phi$ almost unitary and the THD of i_n with a value of 2.2%. It is also worth to note that in all cases the THD is kept below the one demanded in the current related regulation, as the standards IEC-61 000-3-{3,4} and IEEE-519.

Finally, in Figure 7.39, a transition from 49.5 Hz to 50.5 Hz in the voltage source is portrayed. The whole variation has been performed linearly in 40 cycles. It can be noticed that the control system maintains the performance during the complete transition.

For the next experiments, the capacitive-resistive load is connected to the ac source and the network frequency is set at 51 Hz and 49 Hz. Figures 7.40 and 7.41 show that the system performance is similar to the one obtained for the nominal frequency, providing a unitary PF and $\cos\phi$.

7.7 Conclusions

This work showed the architecture, some design issues and a stability analysis for an active filter digital controller based on repetitive control.

The odd-harmonic RC with time varying sampling period designed in Section 7.3 included an adaptive mechanism that follows the time-variation of the electrical network frequency in such a way that the well-known advantages of RC are kept, also maintaining its low computational cost. Theoretical and experimental results proved that the controlled system has a good performance and that, using the frequency adaptation mechanism, it is able to cope with more aggressive frequency changes than the usual ones in electrical distribution networks. However, the variation of the sampling period makes harder both the practical implementation and the theoretical stability analysis.

The controller in Section 7.4, besides the sampling period adaptation, includes an adaptive pre-compensation procedure to follow possible network frequency variations without losing the performance of the repetitive controller. This technique assures inner-loop invariance and, consequently, outer-loop stability, independently of sampling period variations and of the frequency observer dynamics.

Section 7.5 presented a strategy based on the sampling period adaptation showed in previous sections. However, as this operation may negatively active filterfect closed-loop stability, the controller was obtained through a small-gain theorem-based robust control design technique that guarantees BIBO stability in the required sampling period interval, with no restrictions on its rate of change. This interval is set from the expected interval of variation of the network frequency, which is assumed to be known. Theoretical and experimental results showed good performance of the controlled system.

Also, it is worth recalling the relationship between the stabilizing controller (7.28) designed in Subsection 7.5 and the original stabilizing controller (7.13) derived in Section 7.2, which was pointed out in Remark 7.1. This seems to be in accordance with the discussion in [19] for the constant sampling period case, where it is conjectured that the traditional scheme for $G_x(z)$ is quite close to the optimum controller structure. Further research should study the problem in the time-varying sampling period framework.

A new stable second-order IM for HORC has been also proposed and implemented. Although, this repetitive controller achieves a narrower frequency range of operation without reducing its performance, it does not employ neither a frequency observer nor an adaptive mechanism. Finally, the second-order odd-harmonic HORC has similar computational cost compared with the standard full-harmonic RC and the stability analysis of this controller remains in the LTI systems framework. This controller has been validated experimentally in the current loop of the active filter providing a robustness improvement.

Finally, notice that in this work the current reference signal was constructed from the carrier extractor, which yields the voltage source normalized fundamental component. This component could be shifted by an appropriate filtering before constructing the current reference. This procedure would allow the use of the proposed controller architecture for phase shift control purposes. Thus, in upcoming applications it would be possible to provide a controllable phase shift in the fundamental current which may be used to contribute to voltage control in the distribution network.

References

1. Ali, M., Tamura, J., Wu, B.: SMES strategy to minimize frequency fluctuations of wind generator system. In: 34th Annual Conference of IEEE Industrial Electronics, IECON 2008, pp. 3382–3387 (November 2008)
2. Apkarian, P., Adams, R.: Advanced gain-scheduling techniques for uncertain systems. IEEE Transactions on Control Systems Technology 6(1), 21–32 (1998)
3. Bolik, S.: Modelling and Analysis of Variable Speed Wind Turbines with Induction Generator during Grid Fault. Ph.D. Thesis, Aalborg University (2004)
4. Bruijnen, D., van de Molengraft, M., Steinbuch, M.: Efficient IIR notch filter design via multirate filtering targeted at harmonic disturbance rejection. In: Proceedings of the 4th IFAC Symposium on Mechatronic Systems, Heidelberg, Germany, pp. 318–323 (2006)
5. Buso, S., Malesani, L., Mattavelli, P.: Comparison of current control techniques for active filters applications. IEEE Transactions on Industrial Electronics 45, 722–729 (1998)
6. Corasaniti, V., Barbieri, M., Arnera, P., Valla, M.: Hybrid active filter for reactive and harmonics compensation in a distribution network. IEEE Transactions on Industrial Electronics 56(3), 670–677 (2009)
7. Costa-Castelló, R., Griñó, R., Cardoner Parpal, R., Fossas, E.: High-performance control of a single-phase shunt active filter. IEEE Transactions on Control Systems Technology 17(6), 1318–1329 (2009)
8. Datta, M., Senjyu, T., Yona, A., Funabashi, T., Kim, C.-H.: A frequency-control approach by photovoltaic generator in a pv-diesel hybrid power system. IEEE Transactions on Energy Conversion 26(2), 559–571 (2011)
9. Freijedo, F., Doval-Gandoy, J., Lopez, O., Fernandez-Comesana, P., Martinez-Penalver, C.: A signal-processing adaptive algorithm for selective current harmonic cancellation in active power filters. IEEE Transactions on Industrial Electronics 56(8), 2829–2840 (2009)
10. Fujioka, H.: Stability analysis for a class of networked-embedded control systems: A discrete-time approach. In: Proceedings of the American Control Conference, pp. 4997–5002 (2008)
11. Galeani, S., Tarbouriech, S., Turner, M., Zaccarian, L.: A tutorial on modern anti-windup design. European Journal of Control 15, 418–440 (2009)
12. Hippe, P.: Windup in Control: Its Effects and Their Prevention, 1st edn. Advances in Industrial Control. Springer (May 2006)
13. IEC. Electromagnetic compatibility part 2-4: Environment compatibility levels in industrial plants for low-frequency conducted disturbances. IEC 6100-2-4:2002 (2002)
14. Lascu, C., Asiminoaei, L., Boldea, I., Blaabjerg, F.: Frequency response analysis of current controllers for selective harmonic compensation in active power filters. IEEE Transactions on Industrial Electronics 56(2), 337–347 (2009)
15. Mattavelli, P.: A closed-loop selective harmonic compensation for active filters. IEEE Transactions on Industry Applications 37(1), 81–89 (2001)
16. Ryu, Y.S., Longman, R.: Use of anti-reset windup in integral control based learning and repetitive control. In: Proceedings of the IEEE International Conference on Systems, Man, and Cybernetics. Humans, Information and Technology, vol. 3, pp. 2617–2622 (October 1994)
17. Sala, A.: Computer control under time-varying sampling period: An LMI gridding approach. Automatica 41(12), 2077–2082 (2005)
18. Sbarbaro, D., Tomizuka, M., de la Barra, B.L.: Repetitive control system under actuator saturation and windup prevention. Journal of Dynamic Systems, Measurement, and Control 131(4), 044505 (2009)

19. Songschon, S., Longman, R.W.: Comparison of the stability boundary and the frequency response stability condition in learning and repetitive control. International Journal of Applied Mathematics and Computer Science 13(2), 169–177 (2003)
20. Suh, Y.S.: Stability and stabilization of nonuniform sampling systems. Automatica 44(12), 3222–3226 (2008)
21. Teodorescu, R., Liserre, M., Rodríguez, P.: Grid converters for photovoltaic and wind power systems. Adaptive and Learning Systems for Signal Processing, Communications and Control Series. John Wiley & Sons, Ltd. (August 2011)
22. U C T E. Technical paper: Definition of a set of requirements to generating units (2008)
23. Varschavsky, A., Dixon, J., Rotella, M., Moran, L.: Cascaded nine-level inverter for hybrid-series active power filter, using industrial controller. IEEE Transactions on Industrial Electronics 57(8), 2761–2767 (2010)

8
Conclusions

8.1 Conclusions

A wide range of control applications in engineering must deal with periodic signals which usually enter the system as input references or disturbances. At the same time many of these applications operate with periodic signals whose frequency must change by specification or undergoes variations due to the intrinsic features of the specific application or abnormal functioning. In this context, repetitive control has proven to be an efficient technique to deal with constant period signals. Nevertheless, its performance decays importantly when the frequency of the signals is not accurately known or varies with time. This work presents some contributions to the topic of repetitive control working under varying frequency conditions. These contributions are summarized as follows:

- **Stability analysis methods for repetitive control with varying sampling time**. Two stability analysis techniques have been presented: an LMI gridding analysis and a robust control theory approach. Both techniques are useful to verify the stability of repetitive controllers that adapt their sampling period in order to follow period changes in the reference/disturbance signal. The allowable sampling period variation is expressed as an interval. In general, the results show that the obtained interval is larger when using the robust control theory analysis. Although the LMI analysis renders necessary conditions for a sufficient stability condition, it is found that in applications where the varying sampling period is set by a digital clock, the stability can be assured within the interval when the possible sampling periods resulting from the clock resolution are included in the analysis.

    Although several proposals reported in the literature make use of repetitive controllers working with varying sampling period, the stability analysis scenario had not been properly established. Therefore, this results contribute to the stability analysis framework for RC applications that operate with varying sampling period.

- **Design methods of repetitive controllers dealing with varying frequency conditions**. Two designs methodologies aimed at finding a RC which assures performance and stability for a given variation in the frequency of the periodic signals are presented. In both schemes the performance preservation is achieved using the sampling period variation approach. To ensure stability within the given frequency variation interval, a pre-compensation scheme and a robust synthesis approach has been proposed. The pre-compensation scheme assures the stability annihilating the structural changes caused by the sampling period variations through an LTV compensator, thus transforming the LTV stability framework into an LTI one. The robust synthesis design is achieved considering the sampling period variation and the delay element in the IM as norm bounded uncertainties. Using the known frequency interval the norm of the uncertainty due to the sampling period variation can be calculated. As a result, a new stabilizing filter is obtained which assures the stability of the system for the given interval. The obtained controller is similar to the standard design in that the obtained stabilizing filter frequency response may resemble the standard filter one with slight differences. However, the new design method differs form the standard one in that the obtained filter is defined to be proper from design.

These design methods are innovative in the sense they offer a varying sampling period-based strategy for RC that assures the stability and performance within a predefined frequency variation interval of the signals to be tracked or rejected.

- **Robust performance of repetitive control using high order internal models**. Following a different concept, the existing HORC has been modified to obtain an IM intended to deal with periodic signals with only odd-harmonic frequency components. Thus, the odd-harmonic HORC is introduced. The IM design provides action for a wider interval around the odd-harmonic and fundamental frequencies for which the controller has been set, thus providing robustness in the face of frequency changes. As a difference with other methods including the preceding ones in this work, this strategy does not need any frequency observer or estimation.

This design contributes to the field of study in RC in that the odd-harmonic HORC entails an important reduction of the computational burden. This is because the order of the IM is twice as low as the conventional full-harmonic HORC. In fact, the computational load is similar to the one obtained with standard RC, i.e. the standard internal model.

Additionally, the stability of the IM used in HORC has been analyzed. It is found that several internal models may result BIBO unstable. Although this fact does not affect closed-loop stability it is well known that it makes practical implementations harder. For that reason, taking into account the actuator saturation, an AW strategy has been proposed. This is based on a MRAW scheme and the design incorporates an optimal LQ procedure looking for obtaining a deadbeat transition to recover the system. As a result, this AW technique achieves smaller error per period when compared with similar proposals.

Only few AW strategies for RC have been reported. This AW strategy designed for RC is innovative since it provides: 1) a simple and low order design, 2) isolation of RC from the saturation effects avoiding transients and

convergence problems in the IM, 3) a deadbeat transition, which is sought using an optimal LQ procedure that provides a faster transition to recover the system.
- **Experimental validation**. Two experimental applications have been utilized to perform the experimental validation of the presented analysis and control strategie: The Roto-magnet - a mechatronic plant- and an active power filter.

The Roto-magnet application is an educational platform designed to illustrate the IMP in case of periodic disturbances. The purpose of using this plant to validate the approaches in this work is to cover the extensive field of rotating machinery applications where RC can be used. The control goal is to maintain the desired speed by rejecting the periodic disturbances. To obtain varying frequency conditions, varying speed reference profiles were applied in the experiments. The stability analysis and design strategies for RC working under varying sampling period lead to successful results facing varying frequency disturbances. The frequency intervals defined in the design and obtained in the analysis allow a wide range of speed operation. The harmonic components of the disturbance are effectively rejected within the expected speed interval achieving a performance that is almost the same as that of the nominal design. Also, the proposed optimal LQ AW design has been implemented to be used in the HORC case. It is shown that tracking and rejection tasks can benefit from this compensation when actuator saturation occurs. The obtained error is lower compared with other techniques, while the controller control action remains unaffected by the actuator saturation.

The active power filter is an industrial-oriented application. The interest in this field of engineering has been growing importantly due to creation of new demanding needs and applications, e.g. the increase of alternative energy sources connected to the network and new needs in hybrid and electrical transportation. Thus, the control objective in the active power filter application is focused on PF correction and harmonic compensation. The varying frequency conditions are caused by the network behavior and the power factor and harmonic distortion are consequence of the loads connected to the grid. The experimental results show that the RC with varying sampling period adaptation achieves a very high performance despite frequency network changes. The analysis assures stability for the expected operating interval: [45, 55] Hz, while the design strategies fulfil successfully the stability and performance requirements, i.e. PF close to 1 and low THD. Since the load current has in general odd frequency components, an odd-harmonic HORC has been implemented. The experimental results show that this scheme works efficiently maintaining a satisfactory performance for the given frequency interval. The efficiency relies on having lower computational burden compared with full action HORC and on avoiding the use of any frequency observer.

Appendices

A
Implementation of the Stabilizing Filter

Condition 2 of Proposition 2.1 should be fulfilled with an appropriate design of the filter $G_x(z)$. The fundamental issue is to provide enough leading phase to cancel out the phase of $G_o(z)$ [1]. In general, it would be sufficient that $G_x(z)$ approximates the inverse of $G_o(z)$ in the band-pass of the filter $H(z)$ [1, 2].

In case of minimum phase systems $G_x(z)$ is implemented as the inverse of the complementary sensitivity function $G_o(z)$, i.e. $G_x(z) = k_r G_o^{-1}(z)$. The causality issue is solved using the structure of the repetitive controller since $G_x(z)$ is in series with the memory loop element.

Thus, the implementation may follow the scheme in Figure A.1, where r_d stands for the relative degree of $G_o(z)$.

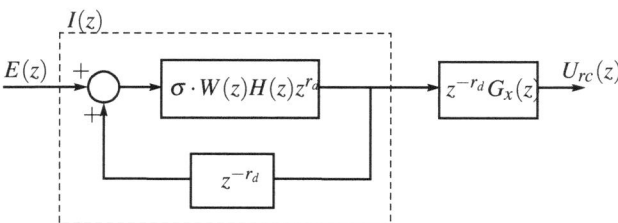

Fig. A.1 Implementation of $G_x(z)$ and the internal model. $W(z)$ is a delay function, $H(z)$ a null-phase low pass filter, and $\sigma \in \{-1,1\}$.

In case of non-minimum phase systems, the design follows the Zero Phase Error Tracking Controller (ZPETC) approach [3] applied to a prototype repetitive controller [4]. Thus, using $G_o(z) = B(z)/A(z)$, where $B(z) = B^+(z)B^-(z)$, with $B^+(z)$ and $B^-(z)$ containing the minimum and non-minimum-phase zeros respectively, $G_x(z)$ is defined as:

$$G_x(z) = k_r \frac{A(z)B^-(z^{-1})}{B^+(z)b_{max}}, \tag{A.1}$$

with $b_{max} = [B^-(1)]^2$; $B^-(z^{-1})$ comes from $B^-(z)$ substituting z for z^{-1}.

Thus, applying this design of $G_x(z)$ to the closed loop function $G_o(z)$ results in:

$$G_x(z)G_o(z) = k_r \frac{B^-(z^{-1})B^-(z)}{b_{max}}.$$

To see how zero phase response is achieved we may express the frequency response of $B^-(z)$ as the sum of its real and imaginary components:

$$B^-(e^{j\omega}) = \text{Re}(w) + j\text{Im}(w).$$

With this, the frequency response of $B^-(z^{-1})$ may be written as follows:

$$B^-(e^{-j\omega}) = \text{Re}(w) - j\text{Im}(w).$$

Then, in the frequency domain we have:

$$B^-(e^{-j\omega})B^-(e^{j\omega}) = \text{Re}(w)^2 + \text{Im}(w)^2,$$

which has zero phase for all ω. As a consequence, the ZPETC (A.1) compensates for the non minimum phase zeros and cancels the phase of $G_o(z)$. Furthermore, the term b_{max} is used to obtain gain close to 1 at low frequencies when $k_r = 1$.

Finally, it can be noted that the implementation also has the structure of Figure A.1 which is used to tackle the causality issue in $G_x(z)$.

References

1. Inoue, T.: Practical repetitive control system design. In: Proceedings of the 29th IEEE Conference on Decision and Control, pp. 1673–1678 (1990)
2. Pipeleers, G., Demeulenaere, B., Sewers, S.: Robust high order repetitive control: Optimal performance trade offs. Automatica 44, 2628–2634 (2008)
3. Tomizuka, M.: Zero phase error tracking algorithm for digital control. Journal of Dynamic Systems, Measurements and Control 109, 65–68 (1987)
4. Tomizuka, M., Tsao, T.-C., Chew, K.-K.: Analysis and synthesis of discrete-time repetitive controllers. Journal of Dynamic Systems, Measurement, and Control 111, 353–358 (1989)

B
Calculation of the Sampling Period Variation Interval

The robust control theory used to analyze the system stability in Section 3.4 yields an interval in which the sampling period variation is allowed while preserving stability. In accordance with Theorem 3.1, the derivation of this interval involves the calculation of the exponential matrix norm $\|\Delta(\theta_k)\|$. Different norm bounds can be utilised to obtain the interval, each one entailing some degree of conservatism. In Section 6.3.3, taking advantage of dealing with a first order plant, a direct calculation has been employed, while in Section 7.3.2, for a second order system, three norm bounds have been applied: a numerical calculation, the log norm of a matrix with respect to the 2-norm [2] and a Schur decomposition-derived bound [4]. This appendix, develops the procedure to obtain the results used in Sections 6.3.3 and 7.3.2 to calculate the allowable sampling period variation interval. The above cited papers have been taken as a reference to describe the procedures. Other bounds and characteristics of the matrix exponential norm are available in [3].

B.1 First-Order Plants

With the stability condition:

$$\|\Delta(\theta_k)\| = \|\Delta(T_k - \bar{T})\| \leq \gamma_{\bar{T}}^{-1}$$

and using

$$\Delta(\theta_k) = \int_0^\theta e^{At} dt = A^{-1}\left(e^{A\theta} - 1\right)$$

for $\theta > 0$, the preceding relation yields:

$$A^{-1}\left(e^{A\theta} - 1\right) \leq \frac{1}{\gamma}.$$

In case of first order plants we obtain:

$$\theta = T_k - \bar{T} \leq \frac{1}{A}\log\left(1 + \frac{A}{\gamma_{\bar{T}}}\right).$$

Finally, we can directly get the upper limit of the interval:

$$T_u = \bar{T} + \frac{1}{A}\log\left(1 + \frac{A}{\gamma_{\bar{T}}}\right).$$

A similar calculation can be done for $\theta < 0$, this yielding

$$\Delta(\theta_k) = \int_\theta^0 e^{At}dt = -A^{-1}\left(e^{A\theta} - 1\right),$$

and, consequently, the lower limit of the interval:

$$T_l = \bar{T} + \frac{1}{A}\log\left(1 - \frac{A}{\gamma_{\bar{T}}}\right).$$

B.2 Higher Order Plants

B.2.1 Numerical Calculation

The numerical calculation procedure consists essentially in obtaining the matrix norm and calculating directly the stability interval. Thus, we have

$$\|\Delta(\theta_k)\| = \|\int_0^\theta e^{At}dt\| = \|A^{-1}(e^{A\theta} - 1)\|;$$

then, to accomplish stability it is required that

$$\|A^{-1}(e^{A\theta} - 1)\| \leq \gamma_{\bar{T}}^{-1}.$$

Hence, we need to find:

$$\theta_{max} = \max\{\theta \geq 0 \mid \|A^{-1}(e^{A\theta} - 1)\| \leq \gamma_{\bar{T}}^{-1}\}$$

and

$$\theta_{min} = \min\{\theta < 0 \mid \|A^{-1}(e^{A\theta} - 1)\| \leq \gamma_{\bar{T}}^{-1}\}.$$

After that, it is possible to calculate $T_{max} = \theta_{max} + \bar{T}$ and $T_{min} = \theta_{min} + \bar{T}$.

B.2.2 Log Norm Bound

Given a matrix $A \in \mathbb{R}^{n \times n}$, the following is obtained [1]:

$$\|e^{At}\| \leq e^{\mu(A)t}, \ \forall \, t \geq 0,$$

with

$$\mu(A) := max\{\mu \mid \mu \in \text{spec } of(A + A^T)/2\},$$

$\mu(A)$ being the log norm of A (associated with the 2-norm), and A^T being the transpose of A.

Using the latter and with $\theta > 0$ we have:

$$\|\Delta(\theta_k)\| \leq \int_0^\theta \|e^{At}\| dt \leq \int_0^\theta e^{\mu(A)t} dt,$$

thus,

$$\|\Delta(\theta_k)\| \leq \int_0^\theta e^{\mu(A)t} dt = \frac{1}{\mu(A)}(e^{\mu(A)\theta} - 1),$$

Then, to meet the stability condition one needs that

$$\frac{1}{\mu(A)}(e^{\mu(A)\theta} - 1) \leq \gamma_{\bar{T}}^{-1}.$$

Thus, the upper limit can be obtained:

$$T_u = \bar{T} + \frac{1}{\mu(A)} \log\left(1 + \frac{\mu(A)}{\gamma_{\bar{T}}}\right).$$

A similar procedure for $\theta < 0$ can be used:

$$\|\Delta(\theta_k)\| \leq \int_0^{-\theta} \|e^{-At}\| dt \leq \int_0^{-\theta} e^{\mu(-A)t} dt,$$

thus,

$$\|\Delta(\theta_k)\| \leq \int_0^{-\theta} e^{\mu(-A)t} dt = \frac{1}{\mu(-A)}(e^{\mu(-A)(-\theta)} - 1),$$

Then, the stability condition is met when

$$\frac{1}{\mu(-A)}(e^{\mu(-A)(-\theta)} - 1) \leq \gamma_{\bar{T}}^{-1}.$$

Therefore, the lower limit can be obtained:

$$T_l = \bar{T} - \frac{1}{\mu(-A)} \log\left(1 + \frac{\mu(-A)}{\gamma_{\bar{T}}}\right).$$

B.2.3 Schur Decomposition-Derived Bound

Given the Schur decomposition of A,

$$Q^T A Q = D + N,$$

with Q being an orthogonal matrix, $D = diag(\lambda_i)$, λ_i being the eigenvalues of A and $N = (n_{ij})$ being a strictly upper triangular matrix, i.e. with $n_{ij} = 0$ for $i \geq j$. We have (see the proof in [3]):

$$\|e^{At}\| \leq e^{\alpha_1 t} \sum_{k=0}^{n-1} \frac{\|Nt\|^k}{k!}, \quad \forall\, t \geq 0,$$

with α_1 being the maximum real part of the eigenvalues of A.

Then, using the definition for $\Delta(\theta_k)$ we can find the following:

$$\|\Delta(\theta_k)\| \leq \int_0^\theta \|e^{At}\| dt \leq \int_0^\theta e^{\alpha_1 t} \sum_{k=0}^{n-1} \frac{\|Nt\|^k}{k!} dt = \Lambda(\theta),$$

therefore

$$\|\Delta(\theta_k)\| \leq \Lambda(\theta),$$

with

$$\Lambda(\theta) = \begin{cases} \sum_{k=0}^{n-1} \|N\|^k \left(-\frac{(-1)^k}{\alpha_1^{k+1}} + \frac{e^{\alpha_1 \theta}}{\alpha_1} \sum_{l=0}^{k} \frac{(-1)^l \theta^{k-l}}{\alpha_1^l (k-l)!} \right), & \text{if } \theta \geq 0,\ \alpha_1 \neq 0 \\ \sum_{k=0}^{n-1} \|N\|^k \left(-\frac{(-1)^k}{\alpha_2^{k+1}} + \frac{e^{\alpha_2 \theta}}{\alpha_2} \sum_{l=0}^{k} \frac{(-1)^l \theta^{k-l}}{\alpha_2^l (k-l)!} \right), & \text{if } \theta < 0,\ \alpha_2 \neq 0 \\ \sum_{k=0}^{n-1} \frac{\|N\|^k}{(k+1)!} |\theta|^{k+1}, & \text{otherwise} \end{cases}$$

where α_2 is the maximum real part of the eigenvalues of $-A$.

Thus we are interested in finding $\theta_{max} = \max\{\theta \geq 0 \mid \Lambda(\theta) \leq \gamma_{\bar{T}}^{-1}\}$ and $\theta_{min} = \min\{\theta < 0 \mid \Lambda(\theta) \leq \gamma_{\bar{T}}^{-1}\}$ and calculate $T_{max} = \theta_{max} + \bar{T}$ and $T_{min} = \theta_{min} + \bar{T}$.

References

1. Dahlquist, G.: Stability and Error Bounds in the Numerical Integration of Ordinary Differential Equations. Kungl. Tekniska Högskolans Handlingar. Almqvist & Wiksells (1959)
2. Fujioka, H.: Stability analysis for a class of networked embedded control systems: A discrete-time approach. In: Proceedings of the American Control Conference, pp. 4997–5002 (2008)
3. Loan, C.V.: The sensitivity of the matrix exponential. SIAM Journal on Numerical Analysis 14(6), 971–981 (1977)
4. Suh, Y.S.: Stability and stabilization of nonuniform sampling systems. Automatica 44(12), 3222–3226 (2008)

C

Optimal LQ Design in LMI Form

The conventional optimal LQ procedure to obtain a state feedback design can be put in terms of an LMI [1, 2]. We are interested in recasting the following problem into an LMI frame:

$$\min_K \sum_{k=0}^{\infty} \chi_k^T Q_p \chi_k$$

subject to:

$$\chi_{k+1} = A\chi_k - B\sigma_{2,k}$$
$$\sigma_{2,k} = K\chi_k,$$

where $Q = Q^T > 0$. Thus, using the Lyapunov function candidate $V(\chi_k) = \chi_k^T P \chi_k$ for the system $\chi_{k+1} = (A - BK)\chi_k$, with $P = P^T > 0$, we have that:

$$V(\chi_{k+1}) - V(\chi_k) = \chi_k^T ((A - BK)^T P(A - BK) - P)\chi_k. \tag{C.1}$$

Using the S−procedure we can state the following:

$$V(\chi_{k+1}) - V(\chi_k) + \chi_k^T Q_p \chi_k < 0, \tag{C.2}$$

this yielding

$$\chi_k^T ((A - BK)^T P(A - BK) - P + Q_p)\chi_k < 0. \tag{C.3}$$

Adding both sides of equation (C.2) from 0 to ∞, one gets that

$$\sum_{k=0}^{\infty} \chi_k^T Q_p \chi_k < -(V(\chi_\infty) - V(\chi_0)),$$

then, assuming for the Lyapunov function candidate that $V(\chi_\infty) = 0$ we have

$$\sum_{k=0}^{\infty} \chi_k^T Q_p \chi_k < \chi_0^T P \chi_0,$$

C Optimal LQ Design in LMI Form

which constitutes the upper bound for the cost function. If we minimize $\chi_0^T P \chi_0$ then the cost function $\sum_{k=0}^{\infty} \chi_k^T Q_p \chi_k$ is minimized and we obtain an equivalent formulation of the LQ problem. Furthermore, the procedure to minimize $\chi_0^T P \chi_0$ can be formulated as a minimization problem in the variable γ that includes the condition $P < \gamma I$.

Now, equation (C.2) can be put in an LMI form. Thus, using equation (C.2), (C.1) we have the following matrix form:

$$-P + \begin{bmatrix} (A-BK)^T & I \end{bmatrix} \begin{bmatrix} P & 0 \\ 0 & Q_p \end{bmatrix} \begin{bmatrix} (A-BK) \\ I \end{bmatrix} < 0,$$

then the Schur–complement formula yields

$$\begin{bmatrix} -P & (A-BK)^T & I \\ (A-BK) & -P^{-1} & 0 \\ I & 0 & -Q_p^{-1} \end{bmatrix} < 0.$$

with an appropriate change of variables we have that

$$\begin{bmatrix} P^{-1} & 0 & 0 \\ 0 & I & 0 \\ 0 & 0 & Q^{-1} \end{bmatrix} \begin{bmatrix} -P & (A-BK)^T & I \\ (A-BK) & -P^{-1} & 0 \\ I & 0 & -Q_p^{-1} \end{bmatrix} \begin{bmatrix} P^{-1} & 0 & 0 \\ 0 & I & 0 \\ 0 & 0 & Q^{-1} \end{bmatrix} < 0,$$

thus resulting in

$$\begin{bmatrix} -P^{-1} & P^{-1}(A-BK)^T & P^{-1}Q_p \\ (A-BK)P^{-1} & -P^{-1} & 0 \\ Q_p P^{-1} & 0 & -Q_p \end{bmatrix} < 0,$$

Then, using $Q = P^{-1}$, $X_1 = KQ$ the following is obtained

$$\begin{bmatrix} -Q & QA^T - X_1^T B^T & QQ_p \\ AQ - BX_1 & -Q & 0 \\ Q_p Q & 0 & -Q_p \end{bmatrix} < 0.$$

Finally $P < \gamma I$ may be expressed as:

$$\begin{bmatrix} \gamma I & I \\ I & -Q \end{bmatrix} > 0.$$

Thus, the LQ design problem can be stated as:

$$\min \gamma$$
s.t.
$$\begin{bmatrix} \gamma I & I \\ I & -Q \end{bmatrix} > 0,$$

$$\begin{bmatrix} -Q & QA^T - X_1^T B^T & QQ_p \\ AQ - BX_1 & -Q & 0 \\ Q_p Q & 0 & -Q_p \end{bmatrix} < 0,$$

where $Q = Q^T > 0$ and $X_1 = KQ$.

References

1. Balandin, D., Kogan, M.: Synthesis of linear quadratic control laws on basis of linear matrix inequalities. Automation and Remote Control 68, 371–385 (2007)
2. Boyd, S., Gahoui, L.E., Ferron, E., Balakrishnan, V.: Linear Matrix Inequalities in System and Control Theory. SIAM Studies in Applied Mathematics, vol. 15. SIAM, Philadelphia (1994)

D
List of Symbols

Repetitive Controller Structure

$H(z)$	Robustness filter
$G_p(z)$	Discrete-time plant model
$G_x(z)$	Stabilizing filter
k_r	Repetitive control gain
$\tilde{G}_x(z)$	Re-designed stabilizing filter
$G_c(z)$	Loop controller
$S(z)$	Closed-loop sensitivity function
$G_o(z)$	Closed-loop system without repetitive controller
$S_o(z)$	Closed-loop sensitivity function without repetitive controller
$S_{Mod}(z)$	Modifying sensitivity function
$S_{Mod}^{hodd}(z)$	Modifying sensitivity function for odd-harmonic high order repetitive control
$I(z)$	Generic internal model
$I_{st}(z)$	Standard internal model
$I_{odd}(z)$	Odd-harmonic internal model
$I_{ho}(z)$	High order repetitive control internal model
$I_{hodd}(z)$	Odd-harmonic high order repetitive control internal model
N	Discrete-time signal period
T_p	Continuous-time signal period
$W(z)$	Delay function
M	Number of delay blocks
w_k	Weights for high order repetitive control
σ	Determines full or odd-harmonic action
T_s	Generic sampling period
$\bar{T}$	Nominal sampling period
T_k	Sampling period at instant k
$\mathcal{T}$	Variation interval of the sampling period
θ_k	Sampling period variation variable
$\Delta(\cdot)$	Uncertainty function

Δ_{T_s}	Sampling period variation seen as an uncertainty
Δ_d	Delay function seen as an uncertainty
$C_{aw}(z)$	Anti-windup compensator
$sat(\cdot)$	Saturation function
r_k	Reference signal
e_k	Error signal
u_k	Control signal
$\hat{u}_k$	Control signal output of the saturation block
$\bar{u}_k$	Control signal input to the saturation block
y_k	System output signal
η_k	Ideal output signal
x_k	State-space variable
(A,B,C,B)	State-space system representation

Electrical variables

i_n	Source current
i_l	Load current
i_f	Inductor current
v_n	Voltage source
L	Inductor
C	Capacitor
r_L	Inductor parasitic resistance
r_C	Capacitor parasitic resistance
$cos\phi$	Current-voltage phase difference
E_C	Energy stored in the capacitors

General symbols

$\mathbb{N}$	Natural numbers
$\mathbb{R}$	Real numbers
$\mathbb{R}^+$	Positive real numbers
$\|\cdot\|$	2-norm
$\|\cdot\|_\infty$	Infinite norm
$\rho(\cdot)$	Spectral radius
$X(s)$	Continuous-time transfer function of system X
$X(z)$	Discrete-time transfer function of system X
$\mathscr{Z}\{X(s)\}$	Z-transform of a continuous-time transfer function
x^*	Steady-state value of the signal $x(t)$
$<x>_0$	DC value, or mean value, of the signal $x(t)$
$<x>_{T_n}$	Mean value of the signal $x(t)$ in a period T_n

Abbreviations

AC	Alternating Current
AW	Anti-Windup
BIBO	Bounded Input Bounded Output

D List of Symbols

DC	Direct Current
DLAW	Direct Linear Anti-Windup
DSP	Digital Signal Processor
FIR	Finite Impulse Filter
HORC	High Order Repetitive Control
IEC	International Electrotechnical Commission
IEEE	Institute of Electrical and Electronics Engineers
ILC	Iterative Learning Control
IM	Internal Model
IMC	Internal Model Control
IMP	Internal Model Principle
LFT	Linear Fractional Transformation
LMI	Linear Matrix Inequality
LTI	Linear Time Invariant
LTV	Linear Time Varying
LQ	Linear Quadratic
MRAW	Model Recovery Anti-Windup
PF	Power Factor
PI	Proportional Integral
PWM	Pulse Width Modulation
RC	Repetitive Control
RMS	Root Medium Squared
SISO	Single Input Single Output
THD	Total Harmonic Distortion

Index

μ-synthesis, 2, 30, 82, 126

adaptive pre-compensation, 31, 86, 119
anti-aliasing, 106

deadbeat, 56, 99, 141
DSP, 114

exponential matrix, 147

Fourier series, 5, 103
frequency observer, 73, 114, 136
frequency variations, 2, 3, 87, 123

harmonic content, 72, 85, 107, 131–134
HORC, 89, 102

internal stability, 31, 33, 87, 123

learning control, 6
LMI gridding, 3, 16, 24, 78, 120, 123, 139
LMI problem, 61, 75
low-pass filter, 7, 72, 75, 90, 107, 129

mean value, 106

notch filter, 106
Nyquist frequency, 6

odd-harmonic, 107

performance degradation, 10, 73, 89, 111
PF, 101
phase cancellation, 9, 145
plug-in scheme, 5, 8, 30, 32, 119

quantification, 74, 114

robust control, 3, 76, 82, 147
Roto-magnet, 68, 89, 99, 141

sampling period, 3, 73, 82, 116, 135
small-gain theorem, 21, 28

THD, 101

MIX
Papier aus verantwortungsvollen Quellen
Paper from responsible sources
FSC® C105338

If you have any concerns about our products,
you can contact us on
ProductSafety@springernature.com

In case Publisher is established outside the EU,
the EU authorized representative is:
**Springer Nature Customer Service Center GmbH
Europaplatz 3, 69115 Heidelberg, Germany**

Printed by Libri Plureos GmbH
in Hamburg, Germany